From Plato to Putin

A Short Guide to the Question of War

By

Andrew Sangster

From Plato to Putin: A Short Guide to the Question of War

By Andrew Sangster

This book first published 2023

Ethics International Press Ltd, UK

British Library Cataloguing in Publication Data

A catalogue record for this book is available from the British Library

Print Book ISBN: 978-1-80441-234-3

eBook ISBN: 978-1-80441-235-0

ACKNOWLEDGMENTS

I am very grateful to the Revd Canon Simon Wright for reading much of this and raising questions and offering suggestions. Also, to the Revd Canon Dr Peter Doll for his support, and not least my wife Carol who read the text and her patience with me for isolating myself as I thought through the problems. I might also add I owe a debt of gratitude to my teachers in the 1960s when studying at King's College London University, men like H. D. Lewis, Eric Mascall and many others who made ethics and moral philosophy meaningful. I am more of an historian than an expert in moral philosophy, but they gave me a grounding which was reliable.

Table of Contents

Foreword

The Revd. Canon Dr Peter Doll

Warfare is a notoriously difficult subject for people to get their minds around. Everyone knows that it brings out the worst in human beings, is destructive of human life and civilisation and wasteful of precious natural and human resources. It engenders cruelty and debases all those who are touched by it. We human beings pride ourselves on our rationality. We know that war is a terrible thing, and yet, generation by generation, human beings continue to choose to make war, continue to dream (contrary to all the evidence) that something lastingly positive can be accomplished by it. It is difficult to come to any conclusion other than that violence is deeply rooted in human nature and is in that sense a 'natural' or 'normal' part of life. Inhumanity is an integral part of being human.

Although warfare is in this sense 'normal', political, and military leaders by-and-large don't want to have the reputation of being warmongers; they look for excuses to explain why they have no choice other than to make war or why the war they want to fight is morally justified. Andrew Sangster brings to this 'Everyman's Guide to War' an unusual but valuable combination of his expertise as a military historian and his moral sensibility as a Christian priest. He registers the importance of understanding the arguments of political and military necessity that lead to war, of recognising the human skill and ingenuity that go into warfare, and of articulating moral despair that human beings cannot break themselves of the instinct or habit of making war.

Because warfare is a constant in human affairs but is also recognised as a moral evil, some of the greatest minds in history have sought to understand why human beings fight, to ensure that wars are begun only for just and moral reasons, and to provide rules of civilised conduct ensuring that war's impact on innocent civilians is minimised. Plato and Aristotle argued that war could be justified on the basis of just political ends, while Thucydides recognised there was only one reason for war – the expansion of power. St Augustine and St Thomas Aquinas formulated what is known as the 'just-war theory', that war could only be a last resort, in order to counteract evil and suffering. Leaders still appeal to these and other ancient authorities to justify their actions and always claim a determination to protect civilian life at all costs, but the reality is that power politics and military necessity almost invariably trump moral restraint, no matter who the combatants may be.

In the brief span of his 'Everyman's Guide', Andrew Sangster makes accessible a remarkable range of original sources and recent scholarship while being always grounded in the nitty-gritty reality of politics and combat. Although he brings the story right up to the present day, he gives us no reason to think today's leaders are any more astute or moral than their predecessors. It may seem remarkable that after 2000 years of Christian history and influence he should conclude that 'the only hope is to try and love our neighbour'; we might reflect on this bearing in mind the saying of G. K. Chesterton, 'The Christian ideal has not been tried and found wanting. It has been found difficult and left untried.'

Preface and Introduction

Having written many history books on 20th century history it struck me one early wet morning while listening to the radio news about the war in divided Syria that 'nothing changes under the sun'.[1] A phrase which epitomises the findings of this exploration. I reflected that there has not been a period in recorded history without conflict and war. I went to my library shelves and started reading from the Oxford Readers publications about war and immersed myself in a variety of books dealing with morality, jurisprudence, and although having a degree in theology and law found the issues complex in the way they were expressed. Many of my history colleagues expressed the same feelings and it dawned on me to write a short book with a style of expression which could be easily grasped, with useful historical events to illustrate the various arguments, thus the title 'everyman's guide'.

My first thought was to explain some situations by comparing an international conflict to a personal domestic situation such as neighbours quarrelling over land ownership, and although this was occasionally a useful device it was not always reliable. When it came to explaining the nature of defensive wars and the use of pre-emptive strikes, wars of prevention, and wars of intervention the only way to illustrate the issue in hand was to refer to actual well-known events from history, and when possible, from recent times.

I felt compelled to keep reflecting on the fact that it is the nature of man which causes war, and this created the first chapter. This caused speculation as to the intrinsic nature and idiosyncrasies of Homo Sapiens, with the initial features of our behaviour as from the time when we first left our caves. The fact that we became the dominant species on our planet against other forms of possible human species and all animals could not be ignored as to why. This of course had to be based on common sense

---

[1] Bible, Ecclesiastes Chapter I, verse 9, (King James Version) 'The thing that hath been, it is that which shall be; and that which is done is that which shall be done: and there is no new thing under the sun'.

speculation but as honest as possible, almost like a spiritual self-examination. This had to be followed by a brief exploration as to how humankind views itself in terms of national identity, as the problem of human behaviour becomes critically important in terms of war at a national and personal level. Individuals are brought up in an atmosphere of what may be called the corporate memory which tends to identify us and our nation and can sometimes dictate our future. The overall history of a nation has many ramifications, but most especially the recent events which pervades the current generation. Nations can sometimes be similar in outlook, but when they differ, it does not lead to living in harmony and war frequently occurs.

War has been the subject of intellectual, religious, and moral discussion since recorded history. It took months of reading the works of early writers which were complex and intellectually challenging, so I took the decision not to avoid their wisdom but to briefly summarise it for the benefit of the reader and not leave a mental fog and sense of boredom, keeping the chapter as short as possible.

The one way to avoid war is to refuse to kill and the idealistic arguments of pacifism are explored and questioned in terms of the pragmatic needs of sheer survival. This is contrasted with what is called political positivism which demands that a balance of power is necessary, and pragmatism or expediency must come before any other consideration. Pacifism demonstrated a high moral stance whereas political realism appears somewhat uncompromising and even cynical. Having introduced two extreme views of war it was necessary to spend a chapter having a brief exploration of the modern era as from 1900 to the current day (2022), not only because context and circumstance have changed and continue to adjust, but the memories of the last 120 years make us what we are today in terms of identity both as nations and individuals.

The next chapter raised the question as to when war can possibly be justified, and the only answer agreed to since the beginning of written records is that of a war of self-defence, despite what pacifists may think. However, this simple argument of self-defence is unsurprisingly not straightforward, as many argue a pre-emptive strike maybe self-defence which is possible, but caution is critical, as it can be used as an excuse for a war of aggression. A war of prevention has been described as defensive, but this is highly suspect as it relies on political speculation. Wars based on intervention for humanitarian causes may have some justification, but there is always the danger of abuse. Finally, the question of terrorism is raised to explore any sense of justification both in its causes and defence against this mode of warfare.

The rules or conventions of conducting war are explored, there are many international agreements on this vexed issue which are often ignored both on the battlefield and at high command level. The central rules of conduct tend to be focused on the treatment of prisoners of war, killing non-combatants and the innocent with the arguments that not all non-combatants may be innocent. The question of torture and suicide missions are examined as well as human-rights. The issues of sieges, blockades, sanctions, hostage-taking, reprisals, guerrilla warfare are all explored alongside the perennial issue of military necessity overriding rules of conduct.

Finally, the postwar scenario is examined and whether it can bring peace and stability after years of destruction and hatred. The various trials such as Nuremberg and others which followed WWII are explored, as well as the South African Truth and Reconciliation scheme as to whether it has anything to offer as a way of hope for the future.

Perhaps the most perplexing and intellectually contentious issue to be explored was that relating to the viability of international law and moral principles being of any value in avoiding war. This chapter concluded that international law is far too lightweight to be of value as it lacks the authority of the municipal law of sovereign states. Morality systems have much to offer but there is no universal morality as there are many different viewpoints in various religions and cultures. The only hope is what is described as natural law which was first indicated by the earliest ancient philosophers that there is an almost inbuilt instinct in humankind that some deeds are simply wrong and unjustified. The problem with this lifeline is that human beings have free will and breaking domestic law is as common as breaking international law and even promises made in treaties and pacts.

The book was originally intended to finish with this chapter, but because of current events the final chapter looks at the issue of today, first at the responsibility of what may be called significant non-combatants who are not so innocent, followed by the sheer number of wars taking place in the last two decades to this day, many not in the public eye. A brief evaluation is surveyed about world tensions, including the dangers of what may be termed as rogues states such as North Korea firing missiles in every direction. There is also a state of nervous anxiety over China's claim in Taiwan, a small island with American support indicating a possible military clash between two superpowers. Finally, the war in Ukraine, the nature of the leaders and reasons for the war, and the various arguments relating to justification mentioned in this study. The inevitable conclusion is somewhat depressing because from whatever angle war and its conduct is explored, war inevitably persists and reoccurs, and rules of conduct are too often ignored. The sadness is that 'nothing changes under the sun' when it comes to humankind's behaviour.

Chapter One - Humankind

There are aspects life on earth which can be disastrous and even challenging to the existence of human life, such as earthquakes, floods, and even an asteroid hitting earth which is often speculated as being the cause for the disappearance of the dinosaur age (Mesozoic period), causing NASA to explode a bomb on an asteroid to see if it could be deflected (26 September 2022). Active research continues to establish how long dinosaurs existed, but it is generally believed their timeline will be millions of years longer than humankind can manage. Climate change is itself a danger which can only be remedied by international cooperation which for many seems farcical, but the other danger is a nuclear war. In 1989-90 when the Cold War ceased it was hoped that this scenario had become an anachronism, but with Putin's war in the Ukraine it is known that fingers are again hovering nervously over the nuclear buttons. It raises the question as to whether there is any point in trying to form legislation to stop war because it is as inevitable and unstoppable as natural disasters.

The question must be asked what is it in human nature which throughout the centuries of our existence constantly drives us to war? When the caveman emerged, like our mammal two legged cousins he probably lived in colonies surviving in small groups. Within each colony there was undoubtedly what has popularly been dubbed the Alpha-Male, one who becomes the dominant leader, who on seeing outsiders coming into the valley felt they were trespassers on their territory and thus emerged the first signs of land ownership. The trespassers would have been ejected or utilised to grow the colony, which community when large enough would occupy the next valley and hills because they offered greener or better hunting grounds.

The writer Yuval Noah Harari in his book on the history of Sapiens indicated that Homo Sapiens was not the only species of what we now call humans.[2] He explained there were at least six different species with

---

[2] Harari, Yuval Noah, *Sapiens, A Brief History of Humankind* (London: Harvill Secker, 2014)

DNA being utilised to track original sources. Whether other species such as Neanderthals or Denisovans merged with our version of Homo Sapiens or not must remain speculation, but over a long period of time Homo Sapiens appeared to become the only human species. Today humankind can be over-sensitive to skin, eye, and hair colour, and it may well be, as Harari suggests, we were the winners in an elimination process. The migratory movements from the African continent continued raising questions as to why Homo Sapiens was successful. One proposal has been that although many animals of all sorts can communicate with one another, Homo Sapiens through a possible brain pattern developed better communication skills, or he may have been more ruthless.

All this must remain curious speculation, but there is something in human nature which is best understood by the swing of a pendulum, a human can rapidly move from being a cooperative kindly soul to being a ruthless killer. From before the time that Cain killed Abel, and from the earliest recorded history of the human species the need for power, dominance, wealth and therefore greed have been hallmarks of humankind, often called 'original sin' by Christians. As will be noted in Chapter Three the ancient Greeks noted this inbuilt propensity in human nature, with Thucydides, Plato, and Aristotle regarding war as originating from humanity. Furthermore, Thucydides thought it so deeply rooted that it could not be prevented or contained.

As a local community grew in power, be it a Greek city-state or a nomadic tribe seeking better pastures and more dominance, these were often the motivating forces for war and the massive migrations across the continents and seas. Seeking resources and better living circumstances, sometimes moving on and eventually settling in a selected spot which became their country led by the more powerful or charismatic warriors of the day. As Michael Howard wrote, war, conflict between political groups 'has been the universal norm in human history', and some societies were more warlike than others, 'for some almost as rite of passage'.[3]

---

[3] Howard, Michael, *The Invention of Peace and the Reinvention of War* (London: Profile Books, 2001) p.1

Leadership was often associated with an overlord or monarchy and supported by other lords claiming noble rights, frequently backed by a religious faith and a military force. From recorded history's earliest times, the settled community in its chosen home of 'milk and honey' was led by the warrior class, namely the origins of the aristocrat, prepared to enlarge its power by conquering neighbours. In Europe much of this was personified by the wandering tribes looking towards the fertile lands of the west, with war acknowledged as an inherent part of human life. They took land, claimed possession by force of arms as there were no property laws. They were not nomadic tribes in the sense of wanderers seeking a living in the deserts of the world, but intent on staying, using agrarian skills, and although Pierre-Joseph Proudhon in his 1840 book on property described 'property as theft' no one would argue with seasoned tribesmen like the Vikings. The territory became their land and needed to be defended from others.

The issue of war was in the hands of the leaders who either obliged or convinced their people to fight for their possessed land, undoubtedly true from the earliest to the most recent of times. War was in the hands of the princes not necessarily the people at large who may have preferred to stay working at home. The Prussian military theorist Karl von Clausewitz after the Prussian defeat at Jena (14 October 1806) by Napoleon noted that 'the people ruled by the Hohenzollern monarchy observed the defeat of the royal troops with indifference', noting the separation of people from their government.[4] Today, if the news is correct, the Russian leader Vladimir Putin in recruiting troops to fight in the Ukraine is facing opposition by many citizens. As the ancient Greek historian Thucydides suggested, the justifications for war are lies that politicians tell their citizens playing on their sense of honour, greed, and fear. In a defensive war the inhabitants often prove more willing, and many might be prepared to fight in the hope of gain, especially seeking a new homeland in better climes, but it starts with the leaders and high command. Even in this modern-day land and resources, be it oil, gas, or water remain an issue.

As such it was the despots and political leaders, and sometimes the military command who initiated aggressive war, leading to the rise and fall of

---

[4] Howard Michael, *Clausewitz, A Very Short Introduction* (Oxford: OUP, 2002) p.18

empires. In the Middle East, Egypt was powerful, then Syria, then the Assyrian empire, the Babylonians, the Persians, the Greeks, and Romans all an onward roll of rising and falling power. In Europe invading tribes grabbed neighbouring lands, fought off intruders, or as with the Vikings some were allowed settlement rather than engage in more conflict. The greed and power seeking of the various types of despots was all part of humankind's nature, and with the passage of time many were consumed by larger neighbours and as with other parts of the world, the Middle East and Far East empires continually rose and fell in the constant see-sawing sway of power.

From the milieu of warring for land based on rights, inheritance, and sheer greed, evolved a pattern of recognised states. There is a generally held historical view that at the conclusion of the notorious bloodthirsty Thirty-Years War the Peace of Westphalia (October 1648) legitimised the state which had its own rights and control of its borders and domestic behaviour. It also produced the pattern of the *ancien régime* of monarch, church, and aristocracy a hallmark for centuries in most European states. Countries differed in their development with France attacking he system in the 1789 Revolution, in England the aristocrats all but destroyed themselves with in-fighting which was followed by landowners and merchants producing a bourgeoise culture. There was always war, either civil or international, the continuing characteristic of humankind not just in Europe but globally.

Historians have in European history given names to the various periods of history, the Dark Ages, Medieval period, the Renaissance, the Enlightenment, the Age of Reason, most of them couched in positive terms. However, in terms of both Eurocentric and global history such is the nature of man the title the 'Pursuit of Power' is more applicable, a title the historian Richard Evans rightly gave to his study of Europe between 1815-1914.[5] This may be regarded by some readers as too cynical, but not a century of recorded history has passed without international conflict, civil wars, and revolutions. There was an era known as the *Pax Romana*, roughly two hundred years when the power of Rome supressed other contenders, but

---

[5] Evans, Richard, *The Pursuit of Power* (London: Penguin Books, 2017)

the migrating tribes still hammered at their doors until the Roman fabric collapsed.

Countries took shape and became recognised states, with systems of governing evolving which were based on varying ideologies based on previous lifestyles, the most common being known as the *ancien régime*, then developing democracies often called the liberal development, various shades of communism, totalitarianism, and some places based their state governance on religious faith. Humankind's characteristic need for power has continued unabated with empire building, colonialism, nationalism. These motives were often a political drive supported by excuses that in occupying another land the aggressors were bringing civilisation or a better religious faith to the conquered. Clausewitz noting that 'war was nothing but the continuation of policy with other means', indicating that 'war cannot be divorced from political life'.[6] Wars persist to this day based on questions of territorial integrity, justifying expansion, religious faith, ideology, mutual safety, often a sense of fear of a neighbour, and many other reasons purporting the justification for attacking our neighbours.

It has been suggested that such is humankind's development that war is as inevitable as natural disasters and plagues, and conflict will never cease, others that some form of international system can 'outlaw' war, but as we progress through the 21st century with the increased dangers of WMD (weapons of mass destruction) such as nuclear power, bio-chemical warfare the stakes have become dangerously high, challenging the existence of humankind and the planet, demanding more attention to the possibility of living at peace in our global community of nations. War is generally acknowledged as dehumanising, albeit a common human activity, but most agree it is evil and leads to unbelievable suffering and many have tried to draw attention to the dangers based on the premise of morality or natural law which will be explored in the penultimate chapter. Clausewitz wrote that war was 'composed of primordial violence, hatred, and enmity, which are to be regarded as a blind natural force' which he could have added seems a characteristic of humankind.[7] This book will try

---

[6] Howard Michael, *Clausewitz*, p.36 and p.52

[7] Howard Michael, *Clausewitz* p.76

and outline and re-awaken the main issues because as man has advanced scientifically and technologically, the inevitability of war and humankind's propensity for playing Russian roulette must be recognised as a danger best avoided.

Chapter Two - Collective National Identity

This book seeks to explore the morality of war, if war can ever be justified and, if so, under what circumstances. As a published historian it seemed right to insert a chapter on the way we remember the past and its wars, and the value and flaws in writing history, as it impacts our current thinking personally and as a nation. A national identity is often based on the collective memory, which is moulded by widespread views of the past gleaned from history books, films, and the media in general. History writing is often open to political manipulation, and not always reliable or easily definable, and constantly varies in analysis. Although a country's long-term history may have some bearing on the collective memory, it undeniably focuses on events of the previous 70-80 years as a generation span. Parents who fought or lived through WWII would impart their views to their children so the generational span can reflect a long period of time.

This British writer, born on the day Auschwitz was liberated by the Soviets remembers playing on the bomb sites in Dover, and can recall the hatred felt against the Germans, the Japanese were never mentioned, and the adults talking about lost friends relating the tragedy and pain of war. It was soon that films of heroism and romantic views of the war were being shown in the cinemas, and it was not until early history books appeared that as a teenager a more informed picture of war was conveyed. Films and journalists were a source of information and emotions, but it felt acceptable to think an historian was always right. A friend of the same generation recalled being informed that Britain had stood alone and won the war, with some belated help from the U.S.A. There was little mention of the Eastern Front, nothing of the Holocaust. Jerries and Krauts were bad guys and Brits were good guys, and he and his friends read a magazine called *RAF Flying Review* which described heroic deeds by people like Douglas Bader and Robert Stanford-Tuck. He also recalled that on looking back his parents wanted to consign WWII to the past, but he also heard about the bombing of some nearby neighbours in the town of Bath and knew his uncle had been a prisoner of war.

When we recall the past, the collective memory of events needs historians who either generated or confirmed the memories of the fading generation who experienced war. Often the historian can question the collective memory by the disclosure of archival evidence and with a degree of hindsight. There is always a constant flow of published literature especially on wars, but extreme caution is demanded when the objective truth is sought. Many historians seek to present their work in this light, but they are only human beings flawed with the influence of their own national background, political standpoint, and possible religious influences. Many histories cannot help but be nationalistic, therefore politically driven or motivated by agendas sometimes not realised by the historian him or herself. History helps provide the collective memory, but sometimes can be controlled by politicians to utilise in their current situation. In the 21st century there are now well over 2,000 years of human conflict to reflect on and draw some conclusions. Just over a 100 years ago this writer had three uncles killed in the Great War, relations who fought in World War II, another in Korea, and friends who battled in Vietnam, Iraq, Afghanistan to mention just a few of the many conflicts within living memory. Each of these conflicts contains its own pandora's box of political debate, interpretations, variations, and moral diatribes. It could be argued that the Great War of 1914-18 was the starting point of conflict, leading to World War II which more than any previous war has touched most of humankind. The historian Patrick Finney in his book *Remembering the Road to World War Two International history, National Identity, Collective Memory* brilliantly drew the reader's attention to the quagmire of historical accounts. In his words WWII 'preserved in western memory as an indubitably "good war" — a status now secured by the enshrining of the Holocaust as its defining atrocity — the conflict also continued to serve as a potent analogical resource'.[8]

History and national identity are crucial because of current national communities sharing a past and seeking a necessary unity in the face of possible future conflict. When a war has finished for the victors, it was a good war, for those who lost a bad war with some memories to be placed

---

[8] Finney, Patrick, *Remembering the Road to World War Two International history, National Identity, Collective Memory* (London & New York: Routledge, 2011) p.3.

on the amnesia shelf. Often the memories are adjusted to suit new developments, which was often the case in postwar 1945 Europe. As the Cold War developed German military figures such as Field Marshal Kesselring who had received a death sentence, found it quickly commuted to life imprisonment and soon released because the Western Powers needed West Germany onside against the perceived communist threat. Views on the past had to be adjusted to suit the ongoing developments in the political world. The writer David Reynolds in exploring Churchill's history of WWII demonstrated how he always had in his mind the emerging potential of the Cold War, suggesting that history often carries political influence between the lines.[9]

Britain

In Britain which survived the onslaught of war and emerged as one of the victors, the popular collective memory was one of success, endurance, heroism, (enhanced by a host of postwar films) but tinged with the knowledge that Britain was a declining world power, with persistent questions over the reasons for the war and why the country had been so poorly prepared. If anything, the war encouraged a deep-rooted xenophobia about continental involvement not helped by Britain as an island with traditional insular notions. There were also deep collective memories that the war may have been against fascism but also the demand for a welfare state after two world wars, and a fairer life for those who fought which explained Churchill's loss in the 1945 election despite being regarded as a national if not international hero. In terms of any collective memory there are varying and frequent conflicting threads.

The nature of British political machinations and the appeasement years has often been a point of memory focus. The collective memory looked back with pride and a hope for a better future, but the question of appeasement in the 1930s remained a serious issue. It raised the question of who was to blame for not responding to what was now regarded as the evident threat by the Nazis. This was epitomised by the article *Guilty Men* written by some

---

[9] Reynolds, David, *In Command of History: Churchill Fighting and Writing the Second World War* (London: Allen Lane, 2004).

Beaverbrook writers, which was a savage attack on those who tried to appease Hitler, who with the benefit of hindsight, was not open to conciliation or agreement. In the light of postwar years, it appeared the government had totally failed to understand what was happening across the English Channel. Postwar, appeasement was initially condemned as a dangerous miscalculation, totally unfathomable and even cowardly. Chamberlain's 'Peace in our Time' claim had initially been met with joy, but was soon regarded with mocking derision, all of which reflected the treasured sense of national identity. The postwar discovery of Hitler's aggressive plans and brutal methods put the appeasement politicians into a defendant's box, especially after the Nuremberg trials. This was first outlined by Churchill's history and appeasement was accepted by many as being dishonourable, purchasing peace at the expense of smaller nations, and in the national memory appeasement for most has become a dirty word, for others a noble cause, offering an example of conflicting corporate memories.

During the 1950s it was evident Britain was in decline, colonies were being returned to their rightful inhabitants, America and the USSR were now superpowers, all of which focused the historical memory on what had gone wrong, rather than accepting that Britain was an off-shore European island. The general public's collective memory may have differed in seeing the return of colonies as a sign of British good will and not a weakness, especially as postwar recovery was improving in their domestic improvements epitomised by Harold Macmillan's 'winds of change' speech in 1960.

One of the strands of the collective memory of the more popular mood, often utilised by politicians, was that Britain went to war to stand by others under threat, which enhanced the belief of Britain's' greatness in adversity and ability to recover. The sense of British tradition, independence and power reflected the image of being the victor (rather than a survivor) in both world wars, and was utilised by Margaret Thatcher in the Falklands War, postulating the British identity as potentially aggressive and winning her the next election. Then Tony Blair took a leading role in the Bosnian-Serbian conflict and Iraq in 2003, and Boris Johnson, saw himself as Churchillian and was the first leader to support and arrive in the Ukraine under threat from the Russian President Putin. How far the collective memory of WWII is pertinent today in the collective memory remains debatable, but it is still invoked by politicians, even during the contentious Brexit debate when the argument Britain once stood alone and succeeded; only time can tell with their use of collective memory whether their appeals were justified.

The influence of Remembrance Day which started in 1919 has grown over the years, it acts as a reminder to the collective memory of the loss of life, but also strengthens the belief that Britain eventually wins its wars, as with the sense of sadness there is always the feeling of triumph. For the current generation another collective memory would be the CND and Aldermaston marches protesting at nuclear weapons, but that has subsided despite the recent friction with Russia over Ukraine. Overall, the British collective memory tends to centre on standing alone whether against the Spanish Armada, Napoleon, or Hitler, and always winning, a dangerous interpretation of the past and too often used by politicians.

America

Following the years of depression and determined isolationism America entered its Second World War after Pearl Harbor, but it had been supporting Britain against Nazism with Lend-Lease, indirect naval support, and following the Atlantic Conference a form of political support. Churchill had badgered Roosevelt knowing that the American economic and military support was critical because after the Great War America had emerged as a major economic power. Following the war's conclusion in 1945, with the knowledge of the Holocaust, the Nazi barbarities in general, as well as the aggressive imperialism of Japan the war was usually described in America as 'the good war'. Many historians and others have spent their time either challenging this concept, including the reasons for entering the European war, and trying to erase the simplification and romancing of what they regarded as the myth. The USA is a huge and diversified country and unlike the European States it is difficult to define any reliable collective memory. The resident American population hardly suffered the bitter experiences of the Europeans as the war was being fought either across the Atlantic or Pacific Oceans. However, it was accepted that the war was being fought against evil tyrannies so moral questions were not raised.

In the postwar years it was more the Cold War and Vietnam which occupied the public conscience. Nevertheless, Roosevelt's constant argument that the 1941-45 war was based on the national essence of personal freedom, elevated in American public opinion that their country was the prime global example of freedom which held firm, becoming almost an American crusade. By the end of the war America had moved from isolationism to global management. There were the usual debates as to why America had intervened but with the emergence of the Cold War, Roosevelt's views still held firm by portraying America as the land supporting freedom. There were, historically, several schools of thought, one seeing the war caused beyond America's frontiers as a threat to their security, and the second that the Axis powers had not been a major threat until American policy became threatening by offering aid to Britain. This also included an attack on Roosevelt for misleading the American people. The former more traditionalist approach tended to survive, using Pearl Harbor as resolving the problem. There were others who defended American isolationism which for them had characterised American principles, and they also criticised the Executive powers for being too far reaching, regarded this power as an attack on the democratic principles. It was also argued that the American mission of imposing their form of morality on bad nations was morally offensive. This continued as America appeared to some to be asserting itself as the prime example for the children of light, for others as asserting itself as the global policeman.

During the 1960s many of these views came under deep scrutiny, especially over such issues civil rights, racial bigotry, Cold War policies, and the start of the Vietnam War resonating in public protests and a deep internal disquiet, especially over American presence abroad. These issues almost sublimated the various shades of a WWII collective memory by challenging the role of American policy overseas. It was felt by some that by going to war against Japan and Germany, America had made itself a prisoner to areas beyond their natural frontiers. This viewpoint seemed to return to the old isolationist argument, that America could do better in passive isolation as an example for others. Many countered by claiming America had to defend the ideals of liberty and shut the door against possible insurgence by foreign ideology or powers. This did not formulate a national identity, but it was orbiting around the issue of what America and Americans should be; it amounted to a search for a national identity especially after the Watergate scandal and Vietnam. President Ronald Reagan a Republican right-winger wanted a return to stability and wealth, and later the collapse of the Soviet Union provided a sense of victory in American eyes. The global mission, often seen by some Americans as central, took another surge of energy from the terrorist attack of 11 September 2001. President George Bush started his attack on terrorists by reasserting America as the global champion of freedom with attacks on Iraq and Afghanistan. It was the old WWII collective memory, but it alarmed many with its signs of xenophobia and military power, and this was not helped by Bush calling it a crusade. When in 2004 the National World War II Memorial was opened in Washington it underlined the continuation of the collective memory of the 'good war'. This was recalling that this war was fought by a united America for moral purposes and for Americans set their identity within the international scheme. America appears to many as a massive country very much divided on many issues in terms of its collective memory. For some citizens America retains a sense of splendid isolationism recently demonstrated by President Trump and his 'America First' campaign. There are many divisions in America, the Right-wing currently portrayed by Trump's followers who see America as the global leaders, and liberal thinking Americans who demand more progress in human rights and racial equality.

For many, still reflecting WWII, there was near universal agreement that Nazism was evil and had to be fought, but it was the start of a long spasmodic war continuing to this day. America at first wanted its way of life adhered to in Europe then globally and this was very much based on the collective memory of the two world wars, giving many but not all Americans the identity of standing as a beacon of freedom to the rest of the world, and to its critics as becoming the planet's police officer.

Russia

The portrayal of history in nationalistic or political terms as part of the collective memory and national identity is a common feature especially to

the main antagonists of WWII. In Russia during the mid-1930s the Foreign
Minister Maxim Litvinov made it clear that Russia was seeking collective
security to stop another war, but the Ribbentrop-Molotov Pact of 23 August
1939 took the world by surprise as two ideological opponents came to an
agreement, not least in dividing Poland. To explain this Russia claimed it
was necessary for the West had failed to join them in collective security.
Such was the eventual embarrassment that the Soviets had to write their
history under the oversight of their political masters. Stalin claimed, by
using Lenin's theory, that the cause of the war was capitalism and its
perverse ways. Later Stalin proposed the thesis that there was much in
common between the fascist and democratic states and allegedly co-
authored a book expounding this theory.[10] It was evidently the re-
structuring the memories of history with a powerful political motive.
Nikita Khrushchev moderated this historical propaganda, but still denied
the secret protocol of invading Poland. Under Leonid Brezhnev capitalism
and imperialism remained the cause of war, with the old claim that
Western policy had not treated collective security seriously. In 1985,
Mikhail Gorbachev tried to moderate the past and end the Soviet fixation
on the Patriotic War, but the historical revisionism was just a momentary
glimpse. The critical political factor was not the Katyń massacre being
avoided, but the secret Protocol over Poland as it had ongoing political
repercussions, and was deeply explored by Western historians, enthralled
by the political ramifications. It was the task of the Soviet historians to add
support to the political demands thereby justifying their national identity.
Vladmir Putin re-invigorated the past by praising to the full the 'defenders
of the Motherland' with constant references to the 'Great Patriotic War'.
Even to this day Soviet texts present the infamous pact and its secret
protocol as being a matter of no choice under the circumstances. Since then,
Putin has encouraged a revivalism of the Soviet era, adapting parts of
Russia's more distant Tsarist past to re-establish the old greatness of its own
form of imperial greatness. It appears with the occupation of the Crimea
and the current attack on Ukraine the defence paranoia on its borders and
sense of a Soviet empire have returned. It would be interesting to know
with certainty how many Russian citizens object to Putin's views and

---

[10] Soviet Information Bureau, *Falsifiers of History* (Historical Information) (London:
Soviet News, 1948)

actions. In terms of this exploration, it is apparent that the collective memory of the past is focused not by the living memory but by current political motives and using the help of official historians to pave the way.

Germany

Nazism in Germany was an aggressive form of nationalism which could be found in various shades in many countries, but Hitler's warning about the annihilation of the Jewish race and its heinous consequences has made the work of German and international historians a minefield. The collective memory and the need to re-discover a national identity presented a confusing kaleidoscope which is outside the scope of this study apart from passing observations. The Germans had suffered from devastating bombing and advancing Soviet troops to find their country occupied and partitioned with an insecure future. Nevertheless, everyday Germans needed an historical explanation to explain their nationwide embarrassment in the global condemnation of the Holocaust and others acts of wickedness within the sinister backdrop of the Nazi legacy.

Historical reasons were diverse and changed at given points during the next 75 years dependent on current circumstances. Some conservative elements tried to remove Germany from Nazism regarding it as an aberration, a time in their history when a tyrant took control, or others explained it as a rupture in German history. It was a time when reeling from WWI the poverty, unemployment which the Versailles Treaty imposed, that some German people felt they were a pariah state and for many Nazism and Hitler appeared to be the only answer as the Weimar democracy, although well structured, had seemingly failed.

In postwar West Germany many regarded the return of a democratic life as successful, but the Nazi past has remained a poignant area of interest for historians and students of politics, invoking a highly pluralistic debate. The painful memories focused on the Holocaust, the involvement of ordinary Germans, controversy over the well-known exhibition of the 'Crimes of the Wehrmacht' (1995) which hitherto had been regarded as fighting a 'clean war' without Nazi taint, and the huge national Holocaust memorial embedded in concrete and the collective memory. Nazism and the people's failure to challenge it soon became another historical issue of debate and divided public opinion. The immediate postwar de-Nazification process only lasted a brief time, because with the emerging Cold War it was decided that a stable society on the Western side of Germany was essential. It was proposed that only a handful of perpetrators were culpable, and most Germans had been victims of Nazi terror, a choice between memory and democracy. This way the German people could blame the Nazi elite, seized upon by German and many international historians. The perceived attempt to blacken Germany's reputation during the Nuremberg Trials could be seen as a uniting call to some historians. Their arguments ranged from restricting guilt to the few, especially Hitler, to a major aberration in an otherwise normal history. There was a bolder historical effort, with some justification, of blaming the Versailles Treaty with its humiliating demands. Others tried to link Stalin with Hitler with wanting to destroy the European balance thereby casting the net of blame on a wider platform. From the 1960s the history of the Holocaust flourished (much prompted by the 1961 Eichmann trial) although the key concepts mentioned above continued in variations. The histories/accounts tended to fall between two schools of thought often dubbed Intentionalist and Functionalist approaches. The former looking towards the leader and his ideological goals, placing Hitler at the centre, the latter that this happened because of structural and economic pressure in a polycratic regime. In the question of continuity, the question was raised whether Hitler inherited the old traditions of the German past, even before 1914.

During the 1980s a degree of nationalism seemed to revive, suggesting the Nazi aggression was not that important because the war was more a matter of being pre-emptive, arising somewhat from anti-Soviet potential stirred by American wishes. Even Operation Barbarossa started to be regarded by a few as pre-emptive because of Stalin's long-term wishes to control Europe, making the Russia leader the principal warmonger.

It is known that many ordinary Germans participated in Nazi criminal activities, others remained indifferent often out of fear or pleased with early victories undoing the Versailles Treaty damage, and some bravely opposed when able. This has led to a German postwar history caught between their suffering and the guilt, a highly complex situation in seeking a new identification for the nation, to which international history has assisted with a conservative thread to the collective memory.

Italy

If in Russia history was controlled by the state, and Germany influenced by having to re-identifying itself as a nation, in Italy there was confusion because of the impact of their ongoing political turbulence. This writer had been reliably informed by a close Italian friend and historian that in his country political patronage is the key to historical success, with diplomatic historians virtually in the employment of the State.

The Italian war years were marked by the figure of Mussolini, the nature of Italian fascism, the switching of sides, the Salò Republic and partisan warfare were all confusing times with on-going political views to this day. In the pre-war years, despite the occupation of Abyssinia and support of Franco, many in the West continued to regard Mussolini as the lesser evil and hoped to detach him from Hitler. To this day the views regarding Mussolini as a person remain critical as many Italians feel their future has been driven by the interpretation of the past. The question often posed was whether he was a mild expansionist, or like Hitler, a severe rupture in their national history which resulted in a horrendous conflict. The past was totally ambiguous, with some claiming the anti-fascist resistance placed Italy alongside the victors because they were the victims of fascism, with others less impressed by the communist resistance. Italian politics has long been a divided scenario ranging from the extreme right and left, although the emerging Cold War influenced Italian political stability for a brief time. The right-wing maintaining a silence while the left-wing, strongly anti-fascist wanted the past to be a permanent memory but both agreeing with the continuous comparison of 'good-Italians' and 'bad-Germans'. For a time there seemed to be a reluctance for historians to indicate any enthusiasm to study fascism. What writing did appear was anti-fascist, typically Benedetto Croce who was a well-known opponent of the regime.[11] He attacked fascism but defined it more as a European invention, a form of

---

[11] See Mack Smith, Denis, *Benedetto Croce: history and politics*, Journal of Contemporary History, vol. 8, no. 1, 1973, pp. 41–61.

aberration which no group had wanted, describing it as corrupt and anti-Italian. It could be regarded by some as an effort to restore Italy's identity yet did not answer the question as to why so many Italians were attracted by what fascism had to offer. This was a rallying call for those on the right-wing that fascist origins were to be found outside Italy, while others purported Italian fascism was another way between socialism and capitalism. Meanwhile the anti-fascists soon took up the expression of describing Mussolini as the Sawdust Caesar, part gangster, part clown, and warmonger. The Italian Socialist and historian Gaetano Salvemini explained Mussolini was never a great statesman, but an irresponsible improviser, half-mad, half-criminal, whose only ability was that of a showman.[12] On the other hand there was support from Luigi Villari who pictured him as an anti-communist nationalist, who Britain and France could have won over providing a balance which may have stopped Hitler in his tracks. There was a degree of shadowing these views among international historians.

By the 1960s, as in Germany, new scholars were appearing and with newly revealed archives more sophisticated arguments were produced, as well as the usual shifts in Italian politics which remained unsettled. This was the time of the Red Brigades and the Left-wing was followed by an anti-fascist turn in the collective memory, but soon followed by suppressing memories of the communist resistance. Each political party held their own readings of the collective memory. By the 1970s the tensions rose with anti-fascist historians stressing the barbarous dangers of the regime with its relationship with Nazism. National identity and its relationship with the past held public attention. The major historian was Renzo De Felice whose major work on Mussolini was deemed by most to be scholarly. The early volumes were seen as acceptable but later the picture of Mussolini changed, unleashing a degree of public anger, as he started to portray Mussolini as a person of some substance, followed by an effort to pass the blame entirely onto German shoulders. This debate ranged between the 'Sawdust Caesar' to a man of insight and a revisionist seeking a fairer world. In Italy (and elsewhere) the debate between a fresh style of nationalism to attacking fascism and its leader simmered on. As late as 1994 Silvio Berlusconi

---

[12] Salvemini, Gaetano, *Prelude to World War II* (London: Gollancz, 1953)

purportedly known for his manipulation and attempted control of historians caused some to feel the taboo on fascist Italy had been broken, and even those who fought for Salò were re-invented as genuine patriots. It was regarded as a means by which the wounds and divisions of the past could be healed. The views of Mussolini received similar cross-sword views amongst international historians. Richard Overy was kinder, Nicholas Farrell described a great man who failed, but MacGregor Knox stressed that fascism meant war. In short it has been a long-term struggle over Italy's collective memory confused by current politics trying to establish their version of Italian identity.

France

There were similar problems regarding collective memory and national identity in France, a country which had regarded itself in the front rank of nations, noted for its civilisation, liberty, a place of culture, intellectualism, and a home for refugees. In 1940 it suffered a humiliating defeat, produced the Vichy state which veered towards collaboration, was tainted by its involvement with the Holocaust, and postwar France lost its colonial importance in Indochina and Algeria, producing a sense of endemic decay.

De Gaulle had represented Free France but during the war years 1940-45 was ensconced in Britain, despised by the Americans, tolerated by the British, and in occupied France the stronger elements of the resistance tended to be communist. As such the French collective memory was as confusing as in Italy. De Gaulle as a politician made a desperate effort to restore the image of France, announcing on the liberation of Paris (the Allies out of political sensitivity standing back as French troops entered Paris) that fighting France, eternal France had liberated herself. The Free French and the Resistance dominated de Gaulle's portrayal of France's recent humiliating history as a phoenix rising from the fire and returning to its grandeur. The immediate postwar collective memory was blurred, confusing, and amnesia was widespread, but 25 years later with a change of generation this started to disintegrate, especially with the Holocaust under intense study, making Vichy's anti-Semitism and collaboration with Nazi Germany more central stage.

There arose a sense of necessity to probe the darker regions of the war years, exploring not just the weakness of the Third Republic prior to the defeat, but the role of Vichy, the Resistance all producing varying interpretations, many of them contentious. Most people accepted the sense of pre-war decadence, but the Vichy and Resistance history caused some serious upheavals in terms of the collective memory. Pétain had once been a hero but after the war became a villain as the blackness of the Vichy regime soon came under the same scrutiny as had the weakness of the Third Republic. Supporters of Vichy blamed the Third Republic and French opinion in the collective memory was as confusing as the Italian kaleidoscope and just as divided. Throughout the Fourth Republic (October 1946) and into the Fifth Republic (September 1958) the French collective memory was fragmented. What was known as the 'Gaullist myth of the resistance' with the airbrushing of collaboration continued for a time, a process de Gaulle considered necessary for the restoration of French national dignity, made more critical with the perceived demands of the Cold War. Histories were not a feature of the immediate postwar years, but there was an outpouring of memoirs by leaders of the Third Republic which were mainly self-justifying, but public memory was more interested in the heroic resistance than the 1930 debacle. When de Gaulle died in 1970 his constructed history of French resistance was soon tested.

When Marcel Ophul produced the documentary film *Le Chagrin et La Pitié* (The Sorrow and the Pity) in 1971 exposing the collaboration, the anti-Semitism, questioning the role of the Resistance, it required some serious consideration from the politicians. President Georges Pompidou the following year suggested it was more appropriate to draw a veil over the past. This was for some acceptable, for others a sense of outrage, as it seemed to produce a sense of insecurity within the national identity. The outstanding French historian Henry Rousso described these latest revelations as the glue which held the French identity together and it was giving way to more critical readings of the immediate past. This was followed by the American historian Robert Paxton who pulled no punches on Vichy collaboration, and was swiftly followed by some French historians, though most were content with criticising the Third Republic, and with a wider perspective of European political machinations. There was a constant flow of revisionist history, but the emotional and political feelings continued to dwell on the years 1940-

1945. France with its traditional left and right-wings with the usual turbulence of French democracy had a powerful right-wing movement (Le Penn), and President Nicolas Sarkozy a right-winger informed the people they had to stop apologising because France was a great country with a great history. In the next century the French were still embroiled in their identity much of it based on the manipulated collective memory, although the melancholy views of Vichy were at last being laid to rest. De Gaulle had established the myths for the sake of French honour, it has taken time for French and international historians to dig deeper, but it led to bewilderment in the French collective memory and complex political involvement, and like many countries the old colonial powers often seem overly concerned about how their country looks on the world stage.

Japan

In postwar Japan, American dominance prevailed, but Japan became a major economic powerhouse and a key American ally in Asia during the Cold War era. The postwar collective memory remained fragmented and divisive. Many leaders regarded themselves as the victims of the horrific bombing raids and the two nuclear attacks sublimating the well-known atrocities of Japanese war behaviour. The atrocities were wide ranging, most prominent was the treatment of POWs created by many films and books on the subject, slave labour, chemical warfare experiments by the infamous Imperial Army Unit 731, enforced prostitution, and the 1937 Nanking massacre. The collective memory of being sufferers and yet creating victims caused a rift not only in the collective memory but in the postwar political scene, even with the ongoing debate over sanitising school history textbooks. One shaft of opinion wanting to draw a veil and another acknowledging the past, while others refused to accept the implications leading to a society and political scene of confusing perspectives. At times the debate seemed almost paranoid if not hysterical, questions of patriotism were raised with the inevitable issue of nationalism. As in Germany some treated the war as an aberration and shifted the blame onto a small group of politicians and military officers.

The origins of the war have proved equally contentious, arguing that Japan's initial war against China was defensive and later to rid Asia of

European and American dominance, to protect themselves from communism, at times pictured as an almost moral campaign. Diplomatic history was utilised to demonstrate the war could have been avoided had diplomacy been better, especially on the American side. In the immediate postwar period, communism was a major issue, there was a purging of the militarists, and a conservative political and economic segment was encouraged. It was far from being a clean purging as in 1957 when Kishi Nobusuke became prime minister because it was known that he had been a suspected war criminal, a time when there was a political swing to the right-wing.

The International Military Trials in the Far East never had the same coverage as in Germany, many feeling they were poorly conducted, with the purpose of placing the war guilt on a few, while absolving the masses who were subjected to the machinations of the chosen guilty. The trial concentrated more on the crime against peace rather than the crime against humanity. There were 11 judges but only three were Asian, and the Indian Judge Radhabinod Pal tended to side with the defendants as he was strongly anti-colonial.

When history books appeared, they reflected the clashes within the collective memory of either justifying the past, the fears of a resurgence of the military class, and became more acrimonious as archival material became public. Outside Japan their attack on Pearl Harbor signified Japanese imperial aggression and tended to remain the orthodox view, though later there were some significantly differing views as the diplomatic past came under scrutiny, and a potentially moderate Japan was given a new slant. As the postwar years unfolded Japan hosted the 1964 Olympic Games signifying that acceptability had arrived. Japan was recognised as part of the Western system and was on an equal footing with other industrial and technological nations, with many old enemies dependent on Japanese trade as customers.

Nevertheless, although conservatism flourished the collective memory, reflecting the national identity continued to twist and turn between questions of being victims and barbaric aggression. As the 1960s, the decade of the Japanese time for the Olympics, the right-wing held sway and there were many articles upholding Japan's right to fight for freedom from outsiders. The Japanese Association of International Relations published essays under the title *The Road to the Pacific War*, some of which tried to shift the responsibility of war away from Japan.[13] Later even some American historians even started to shift the focus onto America as a cause of the conflict, though this was always challenged, but this issue has remained a constant point of contention. Japan had a 'school history problem' and was often challenged by the Chinese and South Koreans, as the sense of war responsibility would not disappear, demanding a stop to airbrushing Japanese crimes out of school textbooks. In 1993 the Prime Minister Hosokawa Morihiro publicly acknowledged Japan's aggressive war, and in 2005 offered an apology to Japan's victims. Ever since the WWII concluded, the collective memory of many in Japan has been confused by the interpretations by historians and politicians of causes and events. There is no singular national identity apart from a natural sense of Japan having restored itself economically, having joined in the community of Western nations, being regarded as important to the Far East, and as the generations pass the conflict of memories and interpretations of what happened may simmer below the surface but no longer dominate, though Hiroshima and Nagasaki will rightfully remain a stark reminder of the danger of war in Japanese and global memories.

Final note

This chapter is a reminder that although most historians remain determined to be objective, they are human and at times flawed, not least because of their national identity and collective memory. Everyone is influenced by political attitudes, ideologies, religious perceptions, and personal experience, which is why the very best historians often deviate from one another in their interpretation of past events. In the case of WWII this is especially true of the causes for the war, the political and diplomatic leaders and then the subsequent 'blame-game'. Historians can lead the way and then journalists with the film media taking up the baton and thereby influencing the collective memory; the media at a mass level carry the

---

[13] Robertson E. M. (Ed.), *The Origins of the Second World War: Historical Interpretations* (London: Macmillan, 1971)

greater coverage and responsibility. As the decades have passed since 1945 there have been new historical insights, fresh debates, and changes in the collective memory often manipulated by the political leaders and climate of the day. There emerges some stereotyping in matters of guilt, suffering, redemption, and in places national conceit amongst the so-called victors. It is always necessary to be aware of historians writing about their own country and foreigners writing of that same country which can sometimes be helpful and leading to contrasting viewpoints. Current situations can influence some historians, the nature of the Cold War, Vietnam, modern terrorism as in the twin-tower attack have a subconscious influence on the best academic minds. During the 1960s there were some seismic shifts in both the collective memory and national identity. There are no fool proof safeguards against political or ideological influences and recognising this makes the reading of history not only safer but more interesting. There remains one area of near total agreement in national collective memories, that war causes mayhem and suffering to a vast degree. This issue begs the question in the study of war raised in the first chapter as to why war starts, and to the next question as to how the best minds in the past have tried to evaluate this problem.

Chapter Three - Brief History of Past Thinkers

Brief introduction

Humankind's propensity to go to war in nearly every generation of recorded history is an undeniable feature of human life, and the histories of war are in abundance. As to the rightness or wrongness of war and why it occurs, there have been some studies which have survived for over two thousand years, but they are often couched in philosophical or theological language and rarely can be regarded as popular reading. They indicate that the question of the morality of war has been studied by a few gifted men who should be noted for the foundations they prepared. This chapter does not give them the time they deserve, but it provides a brief oversight of their various views.

The ancient Greek thinkers (circa 484-322 BC)

In nearly every book or article written about the morality or understanding of war considerable reference is made to the early thinkers who managed to leave their thoughts for future generations. It is known that in ancient Egypt, India, China, and myriad other places various thoughts on war have been expressed, but few have attracted as much attention as the ancient Greeks well-known for their philosophers and historians. They will be briefly noted here simply because they started the process of thinking about the nature of war and its consequences, but their complex arguments will only be noted because it is the writer's concern that this subject must be readable and not bogged down in ancient classical debates. It is sufficient simply to outline their thinking and how they laid foundations for later writers. Aristotle and Plato are two of the most famous Greek philosophers, and Herodotus and Thucydides were ancient historians, the latter was also a general.

The early seeds of a Just War theory as later enunciated by St Augustine of Hippo were hinted at by all four but only by implication with the two historians. They indicated their concerns as to whether a war was justified or not, but they were hampered always by the notion that military leaders in ancient Greece were regarded as noble warriors. Aristotle in his work on *Politics* argued that war's main aim (*telos*–the end result) was to find peace, but later admitted that in preparing for war it helped produced leadership qualities. However, for most of the Greeks in those days war was fought for freedom and loot. Another prevailing attitude was that war was about honour, which has been dubbed the Homeric warrior. The sense of honour was a major concept though it did not prevent the Greeks using the subterfuge of the Trojan horse. Both Aristotle and Plato believed in a soldier's training as having immense benefits claiming it had constructive aspects, but although they, especially Plato, had seen the horrors of war, neither had experienced the devastation and destruction of modern conflict. This made these Greek authors capable of describing their sort of war, being critical as to the reasons for the conflicts they witnessed, but still capable of offering praise for preparation and conduct in battle. As the city states (polis) in ancient Greece grew the nature of war took on a more political significance.

The two historians Herodotus and Thucydides wrote about war, but less on the nature of what was just, whereas Aristotle and Plato wrote on justice, and less about war. Plato suggested that war could be justly fought to allow the city states independence, with Aristotle writing that war could be fought if it had an appropriate aim or goal (*telos*) and pursued this with a proposed ethical system. However, reflecting these ancient times, Aristotle was prepared to see war as necessary for maintaining an appropriate hierarchy which allowed masters to rule over their slaves and keep the barbarians at bay.

Herodotus argued for war to be fought honourably, implying rather than being explicit with the suggestion that war should not be based on dishonourable reasons. Thucydides tended to regard war as originating from a system of power, but both historians were concerned as to how and why war started in the first place. Thucydides looked more to the nature of political and military power and produced the well-known Melian dialogues when the Greeks debated with the islanders of Melos that in the current war their island was needed and because the Greeks were stronger and had the power, they were simply going to take control; it was the ideology of military and national power, ignoring any moral laws. It was to reflect later attitudes called Political Realism which is explored in the next chapter. For Thucydides there was only one reason for war: the preservation or expansion of power.

Much of this debate raged around the ancient conflict of the Peloponnesian war which Thucydides viewed as a tragedy involving the hypocrisy of politicians leading to an unjust war, which as old as these writings are rings bells in the modern era. War was for Thucydides an amoral product for power, which reflected Plato who saw the causes of war as embedded in humankind. However, Plato persisted in arguing that war and justice were bound, and it was a necessary component for training the guardians of the city. Plato's mentor Socrates had noted that the children of craftsmen learnt the trade by watching, and by seeing soldiers fight it would inspire more warriors to unify their city. Some would argue that Plato was the realist because the guardian (legislator, leader) he argued must be prepared to legislate for war because the result (*telos*) must always be peace. Aristotle also argued that war should be fought only to find peace. It has been suggested that Aristotle supported some of the central principles of what would become the Just War Theory, in so far that war is strictly a means to an end, namely the creation of peace. It was a similar base to his ethical system (*Nicomachean Ethics*) when he argued for the doctrine of the mean, avoiding the extremes, and maintaining a sense of balance.

It would take centuries before these ancient Greek thinkers were re-studied, and their debates and views helped formulate later thinking about the nature of war, and as nothing changes under the sun the debate continues to this day. This brief reference to ancient writers demonstrates that the problem of why man makes war has always been questioned by intellectuals and most people.

St Augustine of Hippo (354-430 AD)

When the Christian era opened it proclaimed the Old Testament Jewish decalogue of the ten commandments of which one was 'thou shalt not kill' which seemed to demand an end to war, and had it been universally followed then war would cease, and along with the other commandments there would have been no need for policing. It was, because of man's nature an idealistic Utopia, and stealing, adultery, coveting, killing and all the major sins continued unabated. Killing was considered one of the more serious sins, and in Islam's seven major sins killing an innocent soul is also listed. Built into the Islamic commandment is the qualification that the soul must be innocent, but this is followed swifty by the sin of fleeing from the battlefield unnecessarily.

The Christian demand of what appears to be total pacifism had to be re-set in a context basically to enable survival. The first and most notable attempt was Ambrose's servant and student the well-known St Augustine of Hippo who in his book *De Civitate Dei*, (The City of God) produced the Just War Theory which has been the Christian guide, with variations to this day. In short, Augustine using the text from Paul's epistle to the Romans (Chapter 13:verse 4) explaining God had granted the sword to the authorised government for sound reasons to protect the peace and, if demanded by the government, to punish all wickedness. Augustine argued that if Christians had been in obedience to this command by a wise government and had killed wicked men, they were not guilty of violating the commandment of not killing. Shakespeare in his play Henry V had the soldiers discussing before the battle of Agincourt that they were safe from judgement because it was the king's war.

Augustine's arguments were fluid and appealing, noting that the Romans lived under the same laws as they imposed on others, likening the situation of living in comfort with our neighbours. He was not living in a rarefied ecclesiastical ivory tower as he recognised that a country's governance was sometimes lacking, and peace was uncertain not knowing what their leaders would be thinking of doing the next day. He recognised that all wars, just or unjust created misery, and the language barrier was divisive, but the bloodiest of wars always aimed at peace. It was a pleasant thought that all commanders wanted peace after war, but history has taught us that for some leaders, political and military, occupation of land was the aim, and peace for the occupier could be imposed by strict military law with necessary subservience. He made an appeal to nature pointing out that the wildest of animals desired peace, seeking a mate and raising a family. He argued that fighting a just war was the right thing to do, in defence of one's home and country, and it was lawful to kill sinners based on Christ's statement (St Matthew's Gospel, 5 v30) that if part of our body offends then it needs to be cut off, making capital punishment an option if the argument is taken further. What St Augustine achieved for some Christians was a necessary amendment or guidance for Christians who reacted in defence. It was in many ways common sense because every person knows that if he or she or their family are attacked, they will by instinct defend themselves, and if their lives are threatened, they may well kill the aggressor. Augustine set a standard of behaviour making defence with force a morally justified action. It would be Thomas Aquinas who took Augustine's initial work and built it into a structure which has often formed the basis for justified war for centuries ahead.

St Thomas Aquinas (1225-1274)

Aquinas, a Dominican friar during the 13th century built on the work of the ancient Greeks and St Augustine to form the Just War theory into a long-lasting basis for Western thinking. In his political insights and reasonings in *Summa Theologica* he asserted three main principles. The first that the command for war had to derive from the acknowledged highest leader, the sovereign. This naturally raises the question of whether the sovereign of the nation was capable of being moral in the first place. Secondly, there had to be a just cause or reason to start a war, such as some serious wrong the attackers had done. Finally, those waging war must have a righteous intention such as stopping some evil or promoting a good cause. He also argued that violence must be the last resort and vehemence during the war had to be justified by the circumstances. He focused on the nature of war, questioning the use of ambush and focusing on *Jus in Bello* (the just way the war should be conducted) as well as *Jus ad Bellum* (the justification for going to war). Following the arguments of Augustine, he saw the need to kill if by doing so the whole body is preserved and the common good maintained. As such it was lawful to kill a person who has sinned, but unlawful to commit suicide, but never right to kill the innocent, but to kill in self-defence was acceptable if life were endangered.

Francisco De Vitoria (1483-1546)

Vitoria, who had the courage to criticise the Spanish conquistadors on their conquest of South America, was a Spanish Catholic known as a theologian, philosopher, and a jurist in Spain during the Renaissance era. As Aquinas had expanded Augustine so Vitoria developed in more detail this line of thinking on war and natural law. In the work *De Indis et de Ivre Belli Reflectiones* Vitoria dealt with four issues. First, whether Christians could make war, secondly, where did the authority exist for making war, thirdly, what could be the causes for a Just War, and finally how extreme could be the measures used during a Just War.

They were questions which he answered in depth, relying on earlier thinkers and in brief he argued that self-defence was legitimate but with the provision there had to be a reasonable chance of success. He mentioned the possibility of pre-emptive war against an enemy likely to attack, and the need to punish a guilty enemy. He recognised these major issues could not be covered with wide circumferences and he added further demands. The response to war had to be proportionate with what they faced, and the violence had to be limited to sheer necessary if it could have any claim to be a Just War. He questioned the nature of the leadership able to declare war, arguing that if the citizens opposed the war it would fail as being a Just War, and the people had the right to depose of such a leadership, be it a despot or a government. As with Aquinas he underlined that innocent people should not be hurt and no hostages killed. He may have been writing in the 16th century, but he was relating issues which still occur in the modern world.

He argued that before any war, dialogue was essential with war being the last resort, or as Churchill reportedly once said 'it is better to jaw-jaw, not war-war'. Vitoria argued that if no vengeance was taken on the enemy they might consider another attack, because the end of war must have the aim of establishing peace, which must encompass not just the state but the whole world. He proposed that anyone who killed in a war under the instruction of the government committed no sin, given, he wisely added, only if the law in the first place offered justification in terms of the conscience. This would become an issue in the 1945-6 Nuremberg Trials.

He also added that imperialist wars for taking land, or pillage, or difference of religion was no excuse for conflict nor was the personal glory of the prince who started war. A Just War for Vitoria constantly turned on the question of self-defence or a wrong received. He furthered this argument stating that there was some justification in retaking what had been stolen. This line could involve irredentist nationalism (retaking stolen land) which Hitler utilised to recover territory lost with the Versailles Treaty. He wrote that before any Just War there had to be a careful examination for the reasons. Basing his argument on the Biblical book of Exodus he warned against killing the 'innocent and righteous'.[14] He carried his argument through to the unjustness of spoiling the enemy's future by wrecking his agriculture and means of living, and in doing so encouraged his readers to look to the postwar situation (*Jus Post Bellum*), including the question of slaves of captives.

He concluded with three major canons or rules of warfare, the first being that if a ruler had authority to make war, he should not seek out such opportunities, and should consider his citizens and neighbours. Secondly, when a Just War started it should not set out to ruin the opposition, so peace may ensue, and finally, post-bellum, victory should be underlined with moderation and Christian humility.

Vitoria had, in more detail than hitherto, outlined the Just War standards for the Catholic Church and in the 1992 the Catechism of the Catholic Church reflected Vitoria's and Augustine's thinking for the use of military force.

Hugo Grotius (or Hugo de Groot) (1583-1645)

Grotius was a Dutch theologian, lawyer, diplomat, and humanist who also wrote poetry and plays. His interest for this study was in his book *The Rights of War and Peace*, in which he noted that justice is not part of any war's definition and the two must be treated separately, arguing that war meant depriving another to one's benefit and was ergo repellent to the laws of nature. If it is applied to a person, it followed that it also extended to the

---

[14] Bible, Exodus Chapter 23, verse 7.

action of countries. Individuals often refer to their rights, but if stretched from the private level to that of the state with its claims on people and property it becomes even more significant. The rightness of that right can only be the dictate of pure reason.

From this general statement he moved in the next part of his book to the justifiable cases for war, pointing out that sometimes personal interests are not always looking to justice. Those belonging to justice he defined in three categories, defence, indemnity and punishment. Sometimes it is caused by fear of one's neighbour and Grotius argued that this apprehension was insufficient in terms of justice, because all these issues demanded was a strong defence. Otherwise, it would be little better than preferring the neighbouring lands or reducing another country to a state of servitude. He also noted that the initiation of war may have been justified, but during the war it can become unjust. It is always better when lives are in danger to err on the side of safety even if it may prove detrimental, but better this than becoming unjust. In terms of supporting a neighbour in distress it does not necessarily compel reasons for taking to arms, and the sovereign's first responsibility is to the citizens. However, in terms of treaties it is a matter of common humanity to defend others, but always taking care that one's own state must be preserved.

As the student works through the history of past debates and views it may seem at times repetitive, but each thinker always appears to add a new dimension or insight. The reason is probably the changing of times, new developments in the reasons for war, changes in weapons and the conduct of war, as well as disputes as to handling the postwar situation.

Samuel Pufendorf (1632-1694)

Samuel Pufendorf was a political philosopher, jurist, economist, and historian and for this study his work *De Jure Naturae et Gentium Libri Octo* (The Laws of Nature and People, Book Eight) written in 1672, added his ideas to those of Grotius and others to develop laws in common for nations. He argued that there is a state of peace with men, but it is constantly fragile, and the peace of a state depended on the will of the individuals within that state, which was contingent on reason based on natural law. He also made the critical point that any form of international law should not be limited to Christian nations but be a link between all peoples.

He claimed natural law (which will be explored in Chapter Nine) demanded no one unjustly hurts or damages another, pointing out that even animals depend on instinct to survive by being defensive, pointing out that war can sometimes be necessary when an evil threatens, but he proposed there should be strict limits and sound reasons for man to make war. He limited the causes of war to self-preservation, protection, and the causes for war should be clearly understood. He thereby classified wars into defensive or offensive with the latter only being initiated by the need for defence. He referred to Grotius and his views on unjust wars of which he was at times critical. Ambition and imperial greed were for Pufendorf self-evident, but the fear of a neighbour does not always constitute the basis for war unless it can be proved beyond doubt that the neighbour is intent on an attack. He also noted that it was wrong to use a greater evil to defend oneself, and that all must be kept in proportion. He added to the debate of coming to the aid of another nation the original cause of the war must be lawful, and in offering aid a country must look to its citizens first, as folly must be avoided at all costs.

Thomas Paine (1737-1809)

There is an endless list of men who engaged their intellects on the question of man and war. Thomas Paine in his views in *The Rights of Man* became central to much thinking. He argued that man could live in harmony without the vested interests of governments and had the right to revolt when a government did not protect the natural rights of its citizens. He stated need for peace to establish democratic institutions, but this needed free nations who would talk through a problem rather than resort to war. This was an ideal world Paine portrayed, almost a Utopia unlikely to be achieved.

Jeremy Bentham (1748-1832)

Bentham was the modern founder of Utilitarianism, proposing the greatest happiness for the greatest number, argued for individual and economic freedom, including equal rights for the sexes. He wrote *Plan for a Universal and Perpetual Peace* in 1789, on the eve of the French Revolution. He proposed that if it were not for colonial problems Europe would be at peace and is often regarded as the founder of British liberal thinking, looking to a Common Court of Judicature to settle differences between nations, with the demand for a free press. He argued that a country could occupy adjacent land if it were not being used, an argument which has been exploited by many leaders and governments.

Conclusion

In making this brief survey of previous thinkers from over 2,000 years ago to the 18th century the main thrust in the overall developing line of thinking continued to centre on the Augustinian theory of the Just War as being a matter of defence. Going to war for reasons of imperial ambition, territory, glory, political dominance, fear, religious faith, are denounced as the centuries pass. Also studied were issues such as aiding a neighbour, proportionality of action, conduct in war, pre-emptive or even preventative war as a defence measure arise, and punishing evil doers in the postwar situation. The concept of some form of international law was raised, but as with all matters relating to the questions of the morality of war it was studied by philosophers, theologians, and a variety of other international dedicated thinkers, but played little part in the discourse of the wider world. It was rarely discussed beyond the confines of academia, rarely a subject in the barracks, the fields, the workshops, and by politicians only when it suited them and the demands of the day. As man moved into the modern era the dangers of war increased as technology grew and then science made the destructive force of weapons greater, and finally with WMD the threat of global annihilation, but the question of morality tends to remain on a library shelf...such is the nature of man.

Chapter Four - Unlikely and Unattractive

Introduction

This chapter will explore two aspects in approaching war. First the subject of Pacifism which is exemplary in its demand for high moral conduct, but it raises major questions as to its practical viability given man's propensity for aggression. The second part is a stark contrast as it explores Political Realism which purports to seek power to establish a balance of power and therefore peace. It is based on political drive, which is prepared to ignore other schools of thought, be they economics, morality, religious faith, if political necessity demands otherwise. Political Realism may appear at times ruthless, but it has often been deployed sometimes openly and other times 'under the counter'.

Pacifism

There are no 'ifs and buts' about the Ten Commandments, they are written as imperatives and treated as such by some people. They demand a moral standard which is beyond most human beings and sometimes need to be challenged on ethical grounds. Imagine an armed maniacal gunman coming into a classroom and asking the teacher to point out a pupil because he had to be killed. Does the teacher obey the commandment not to lie and point out where the victim is sitting, or lie and say he never came to school that day? If one's family is dying from starvation does a person not steal a potato from the wealthy farmer's field as a last resort with some justification? Worse still, imagine a battle scenario where your friend has been fatally wounded, he is conscious, in appalling pain, there are no medics anywhere, no morphine, no aid and not any likelihood of help and he begs you to 'finish him off'. Do you obey the commandment not to kill and leave him to suffer, or do you go behind him and shoot through the head to stop his agony. Do we not defend ourselves and family in the face of a lethal attack? Even Christ challenged the commandment not to labour on the Sabbath, asking people whether they would not rescue their ox from the ditch on the Sabbath.[15]

---

[15] Bible, Luke's Gospel, Chapter 14:5.

The absolutist stance of some of the Jewish and Christian commandments becomes questionable in times of emergency, especially war. As noted in the previous chapter the Christian viewpoints had to be made more flexible because although war was regarded as evil, it could not be avoided and had to be faced if survival were at stake. As such Christianity had to change its ethical rules but it 'became, and remained for a thousand years, one of the great warrior religions of mankind'.[16] Even during the period of colonisation of other countries by European powers it was often claimed occupation was for the benefit of the so-called savages to bring them Christianity and civilisation. This did not always ring true with all Christians, but the Church often supported the state and even had interests in the slave trade.

Despite these reasonings of defence and making a better world, there were many who thought that killing under all circumstances was immoral and the only true way to avoid war and killing was to be a pacifist. Perhaps the most influential historical writer on this subject was Desiderius Erasmus (1466-1536), a Dutch Philosopher and Catholic theologian. He wrote that war disgusted him, he despised the military with its so-called profession of arms, and he supported the claims of early Christianity claiming humankind was at fault. He started what is often termed liberal pacifism, stating that rulers would be better occupied looking after the welfare of their citizens and their country, and not be pre-occupied with taking their neighbour's land and causing suffering for untold thousands. He was writing at a time when wars of aggression were common, during the time of Pope Julius II who was known as sponsoring great art works in the Italian renaissance but was also called the Warrior Pope, having chosen his name from Julius Caesar and described by Niccolo Machiavelli as the ideal prince. It was a time when Erasmus's dealing with the brutality of war was made necessary by the current climate of the day. Erasmus never had to deal with the complications of rule and governance unlike his friend Thomas More (1478-1535) who during the time of Henry VIII was committed to political activity. In his famous work Utopia, the citizens of his ideal land hated war, which they approached with extreme caution, but

---

[16] Howard, Michael, *War and the Liberal Conscience* (London: Hurst & Company, 1981) p.5.

Thomas More held to the Augustinian view that defence of territory remained essential, and he allowed heretics to be burnt at the stake to save their souls.

Erasmus held to the pro-active pacifist line, but it has been suggested he was not totally against war as in *His Instructions to a Christian Prince* he thought war against an infidel could be justified, but he always blamed the princes for having selfish reasons to provoke war; nevertheless, he was the standard-bearer for pacifism. He persistently pointed out Christ's demand for peace and understanding, but that humankind's mind was constantly focused on lust, wealth, power, and revenge. Erasmus pointed out that wild animals had teeth and claws, but did not group together to wage war, which he designated as a nasty human streak. Even if it cost lives and money, peace he argued was the only alternative, and even the infidel could be won over by friendly approaches. It was his main belief that Christians had to unite to abolish war.

Another cleric, the French Monk Eméric Cruce (1590-1648) opposed war like Erasmus seeing no justness in any sort of war, but he proposed an international body to disentangle international problems. Not only was he proposing a form of the United Nations hundreds of years earlier than 1945-6, but also argued that society's structure had to be changed to avert war. He proposed that military forces should be reduced simply to act against pirates suggesting a form of police force rather than a full military establishment. He foreshadowed future thinkers such as Adam Smith, Richard Cobden, John Stuart Mill in suggesting that economic development would establish peace, which was another good idea which never worked because the need for economic power often created war.

The Society of Friends or Quakers founded in the mid-17th century, the time of the English Civil War, becoming the notable Christian movement for pacifism. The first Peace Movements were founded on the grounds of the absolute pacifism of the Christian Gospels and spread abroad, especially in America. War was denounced as pure evil and even arose in political circles, and in 1816 the Peace Society was founded in London for the Promotion of Permanent and Universal Peace. There followed a series of Peace Conventions in major cities such as London and Brussels and there were many Peace Congresses. Men like Cobden managed to link the movement to the Free Trade elements on the grounds that economics and trade took away the reasons for war.

However, it is generally accepted that the Crimean Conflict (1854) hindered any progress because the public appeared enthusiastic about war and defeating the Russians. Since the wars of Napoleon men like Wellington and Nelson had been applauded as national heroes, even though Cobden had argued that the aristocracy had cleverly worked on the minds of the public including the working-classes. The social elevation of the military has been described as the Prussian disease, but other forces were now at work, with a belligerent form of nationalism developing which seemed to unite the leaders with the citizens. Extreme nationalism, as opposed to patriotism, had a marked effect on public opinion, where even the belief that people were prepared to sacrifice themselves for their country. It put the pacifist into an unpopular role, and the famous French pacifist Jean Jaurès was assassinated in a Parisian café by a 29-year-old nationalist called Raoul Villain, on the grounds that pacifism was an act of treason.

Nationalism was on the rise across Europe and elsewhere on the globe, and it appeared that it was no longer the princes of kingdoms starting wars for glory and power, but now whole nations with their various forms of government seemed prepared to make war. It had a major impact at the start of the 20th century with the Great War 1914-18, with women singing recruiting songs with lines such as:

> We've watched you playing cricket and every kind of game
> At football, golf and polo, You men have made your name,
> But now your country calls you, to play your part in war,
> And no matter what befalls you, we shall love you all the more,
> So come and join the forces as your fathers did before.

White feathers were given to those who refused to fight implying they were cowards, the term CO, standing for conscientious objector became part of everyday debate, and pacifism struggled to be popular in the public eye. The war had become a highly emotional and contentious issue, not based on its causes, but on the popular nationalistic upsurge. The carnage of the

1914-18 war with the loss of a generation soon raised doubts and poets such as Wilfred Owen increased the sense of cynicism in the call for dying for one's country.

Religious faith had its differences, but many faiths tended to back their political state. When the U.S.A. entered the Great War in 1917, a Cardinal James Gibbons told all Catholics to support the war on the grounds that Our Lord Jesus Christ does not stand for peace at any price, and that however well-intentioned pacifism maybe it was mistaken. In the same way that the current Russian Orthodox Church appears to be supporting President Putin in his war against Ukraine.

Nevertheless, the moral debate to end war by pacifism still had strong intellectual and respected backers in the form of Leo Tolstoy (1828-1910) and Bertrand Russell (1882-1970). Tolstoy used powerful arguments and never hesitated in denouncing human folly, arrogance, greed, and ambition. He argued that even self-defence was no argument and although it was understood that war could offer only dire consequences, he argued it was impossible to know what not-fighting would produce, because it had never been tried and therefore impossible to know. The year before he died Tolstoy wrote for the 18th International Peace Congress in Berlin (1909). He noted the old argument that millions could die on the orders of a few men, the governments who depended on armies, but it was impossible to ignore the command of loving one's neighbour, otherwise it was a criminal act in the eyes of God which, he claimed, was the collective wisdom for Christian nations.

This was written prior to the Great War but Bertrand Russell wrote with the knowledge of the carnage it caused.[17] In arguing against fighting he believed fewer lives would have been lost through passive resistance, he postulated a Council of Powers, suggested that if England had disbanded its military the Germans would have had no reason to invade, and if they did there might be changes, even a few executed, acknowledging that it would take courage to hold strikes and offer passive resistance. He further argued that the Germans could have the colonies because Russell noted, we

---

[17] Russell, Bertrand Vol 13 (ed. by Rempel, Richard A,) *From Prophecy and Dissent*, The Collected Papers of Bertrand Russell (London: Unwin House Hyman, 1988)

fear the external enemy who kills the body, but more serious is that which kills the soul. In this diatribe Russell is making huge assertions, that if an enemy walked through open doors, he would not enslave the citizens and exterminate others. He may have been morally right but whether his views were sustainable is another question. He goes by what he deemed right or wrong but offered no consequences. For most people it is a dream to have and hold peace and to remedy the humankind propensity for power and influence, but it raises the question as to whether pacifism is a form of surrender to a potentially dangerous future.

Many hoped the League of Nations could establish peace, offer a collective security but along with the United Nations (hereafter UN) neither have managed to stop war happening, an unpleasant reality explored in Chapter Nine. It was the hope of some form of collective security which inspired the famous Oxford Union debate (1933) on the proposal that the 'House would not fight for King and Country'. It was a moral stance of the highest order, and pacifism is moral, it is an absolute creed, but in most people's views it is untenable because of the nature of humankind. Many of the students in that famous debate were soon fighting the Nazi aggressor within a decade. Wars have started through the need for glory, power, land, wealth, fear, between tribes, then princes, between states, and added to the kaleidoscope of reasons the last century has added to the witch's brew ideologies such as totalitarianism, communism, various shades of democracy, nationalism, and the inter-faith tensions persist. Many conscientious objectors showed considerable courage in both World Wars not just by standing up for their beliefs but often working on the front lines in capacities as medics, drivers and so forth. The film *Hacksaw Ridge* (2016) showed the life of one such objector, a Desmond Doss, who would not fight but won awards for bravery in saving lives. Each person has a view, but it should not condemn pacifism even if in most peoples' views it is untenable, because even the most peaceful of animals will fight to defend their family, because sadly, in human life, there will always be an aggressor and self-defence makes common sense.

Political realism

At the opposite end of the spectrum of pacifism is what has been called political realism which offers a pessimistic view of human nature by arguing that only the facts and pure reason should be the central concern. This was best expressed by Clausewitz when he wrote 'the fact that slaughter is a horrifying spectacle must make us take war more seriously, but not provide an excuse for gradually blunting our swords in the name of humanity. Sooner or later, someone will come along with a sharp sword and hack off our arms'.[18] Whatever views there are of Clausewitz, his cynical approach about mankind has some justification as every decade in recorded history there have been serious wars or minor conflicts.

This argument for Positive Realism was reflected by the Hans Joachim Morgenthau (1904-1980) who was a German American jurist and political scientist. He claimed the globe was in a continual anarchical state for which the realist demands military strength, because self-preservation must rule all decisions. The realist argument is old and was portrayed in Thucydides account mentioned in Chapter Three when the Greeks told the islanders of Melos that they would be occupied because Greece was more powerful and therefore it would happen. It is noteworthy that Machiavelli advised prudence before morality. Political realism was seen by some in the Cuban crisis (October 1962) where the odds were weighed up politically and militarily and, it is claimed, avoided total war.

Morgenthau based his argument on six fundamental points. It was a pessimistic view of humankind because it was based on the sheer lust for power, accused Liberalism of failing, with the selfish nature of humans as its central theme. The argument evolved around political realism. His first point was that society and politics are controlled by objective laws which contain their roots within human nature, that there must be a clear distinction between the reality of the truth and mere opinion and having noted the essential facts, apply pure reasoning.

Secondly, he argued that political realism is the only matter of interest defined in the need for power. If this is accepted and understood then rationality can work, and it should not be confused by concerns over motive and personal ideology, implying that only this approach can offer a sound foreign policy and an appropriate balance of power. It implied that

---

[18] See Howard Michael, *Clausewitz, A Very Short Introduction* (Oxford: OUP, 2002) p.47

although a politician's motives may be based on ethical principles it was not necessarily best for the country. In the late 1930s the Western efforts to seek peace with Nazism may have been ethical but it failed and led to war and suffering, whereas going to war earlier may have destroyed Nazi aggression before it became too powerful. It is argued that the political machine should deal with the facts and not the ideals and hopes.

Thirdly, political reason assumes, through what it addresses as sheer reasoning, that existing matters may change, and a key objective of the original reasoning cannot remain static as the context may vary. A one-time aggressor may suddenly become an ally, or a new approach may be revealed, and 'sticking to one's guns' could be irrational under changing circumstances. This was epitomised during WWII when Stalin having joined with Hitler to invade Poland was then invaded by German forces and became an ally to the Western powers.

Fourthly, political realism would naturally be aware of moral and other political tensions within a situation, but such concerns must come under the scrutiny of the known and concrete situation as of that time and not rely on any form of abstract reasoning. This line of thinking may have reflected the times when dropping atomic bombs on Japan occurred, and the knowledge that the Western partner was Stalin who might become an ideological threat.

Fifthly, political realism must not fall into the trap of taking into consideration the moral guidelines of another country compared to the moral laws which govern everyone else. This again suggests the possibility that the Katyń massacre, known to be Russian, was not raised except as a German atrocity. Others should be evaluated by the same rules that one's own country follows. This naturally assumes that 'we' have all the answers to what is right or wrong and 'we' know what is moral and has led to problems in the so-called recent wars of terrorism.

Finally, the admission that there are notable differences between political realism and other areas of thinking, but political realism demands an interest only in power in terms of the various understandings of humankind and its propensities. It becomes critical that the political realist must clearly indicate where the nation's interests clash with the moral and legal viewpoints. This line of thinking elevates politics above all other schools of thought and concentrates on the politics ensuring the appropriate balance of power despite anything else. After WWII the moral turpitudes were prominent in most peoples' minds and there was an inbuilt distrust of any form of totalitarianism as seen in Hitler and Mussolini. President Truman never changed his mind about Franco the despised Spanish dictator, but his military men, especially General Omar Bradley were now showing considerable interest in Spain, and this included Dwight Eisenhower the NATO Commander-in-Chief for strategic reasons.[19] In June 1951 they made contact with Francisco Franco with a view to using Spain for military bases. The Americans were continuously aware of the European distaste for the Spanish dictator, but 'concluded that military necessity outweighed political sentiment'.[20] Franco was thrilled for not being treated as a pariah and made much political show of the agreement, but for the Americans political realism was the pressing issue as the Cold War developed.

In his work *Politics Among Nations*, Morgenthau proposed that skilled political diplomacy based on political realism could bring stability and peace.[21] It followed that the nation attempting this must be the superior one in power. It was perhaps not surprising giving that America regarding itself as a beacon of light to the free world became for the more cynically inclined the need to be the world policeman based on its military power. Morgenthau supported the Roosevelt and Truman administrations, and he became a consultant to the Kennedy administration, but parted company in the time of President Johnson over the Vietnam War which he considered a mere civil war with no global ramifications, perhaps with some justification.

Basically, the arguments for political realism remove any moral understanding unless it suits the politicians and basically denudes

---

[19] Omar Bradley saw the peninsula as "the last foothold in continental Europe." See Payne Stanley & Jesús Palacios, *Franco, A Personal and Political Biography* (London: University of Wisconsin Press, 2014) p.311.

[20] Preston Paul, *Franco* (London: Fontana Press, 1995) p.612.

[21] See Morgenthau Hans, J, *Politics Among nations: the Struggle for Power and Peace* (New York: Knopf (4th Edition), 1967)

humankind of any sense of moral propriety or even civilised behaviour. Political Realism leaves the control of the globe's future in the hands of politicians who can act without reference to any of man's accepted moral guidance systems, ignore if deemed necessary any national ethical conduct, cast aside other schools of thought, and simply seek power on the pretence of maintaining stability and peace. The question must be asked whether politicians can be trusted, and whether political realism is overly cynical when trying to preserve a peaceful coexistence between states.

Chapter Five - The Modern Era

Historical overview of the modern era

Wars have started for multitudinous reasons, but mainly based on territory, wealth, and power. Just less than 300 years ago there was war between England and Spain (1739-48) which was basically a colonial war caused by British and Spanish commerce and the British slave trade. A British captain of a smuggling merchant vessel called Robert Jenkins had his ear cut off by a Spanish coast guard in 1731. This incident was later used to stir up outrage at Spain and the war was later dubbed by the historian Thomas Carlyle as the War of Jenkin's Ear. It had nothing to do with a lost ear, but this expression certainly underlined the all too frequent excuses for war breaking out.

Forty years after this war ceased the 1789 French Revolution created a major European war which purportedly set out to liberate nations from their form of the *ancien régime* from which the French had hoped they had liberated themselves. The revolutionary French regarded themselves as the standard bearers of liberty but were seen as the aggressors across Europe with a major international conflict which introduced guerrilla warfare. During the postwar period in the 19th century nations started to seek their own national identity, in Germany the *Volk*, which from a number of small states produced Germany. The same happened with the of unification in Italy, unifying small states and feudal titles making one state. The scene in Europe was transforming from small states into major powers. A new world order emerged based on a growing sense of nationalism, liberal views were surfacing, the traditional conservatives never left the scene, and the balance of power seemed to hold Europe steady for most of the century. People like the liberal thinker Richard Cobden claimed international commerce would bring peace, the conservatives that the existing order was working, and nationalists with their belief that states had the right to fight for their existence if not expansion. During the early part of the 19th century there was a sense of some peace, however, tensions remained between the French and Austrian powers with Britain being insular showing little interest apart from colonising various areas around the globe. There was the rapid growth of industrialisation and literacy improved as the working class evolved in the new era. The new European order creaked, and various causes have been suggested. First was the growing nationalism encouraged by governments, in Britain maps of the world coloured the empire in pink, military power was increased with new technology, and it was not long before nationalism looked to military forces to impose their concept that their nation was first and foremost.

Nevertheless, attempts were made to build an international body of common will, the Hague Conferences of 1899 and 1907 with the founding of an International Court of Justice, but nationalism which reflected a base tribalism remained all too prominent. War was seen by the misrepresentation of Darwin's natural order of things, and there 'would be a distorting of reality to pretend that it was stronger in Wilhelmine Germany than elsewhere in Europe'.[22]

The Great War 1914-18 witnessed the popular appeal of nationalism with America joining in the European war and emerging as a world power. With the Russian Revolution there arose new ideologies of communism along with fascism in Italy, Germany, and Spain, liberal democracies, and much later in the century the return of religious extremist fundamentalists. These developments provided even more causes for war. Lenin had argued that war was a tool by the ruling classes to hold down the oppressed, then necessary for the desire to convert the world to communism, and the possible causes of war which had hitherto been many, increased.

The Paris Peace Conference and the League of Nations (1920) offered some hope, but the Treaty of Versailles by humiliating Germany and making it a pariah nation despite its inherent strength paved the way for more war. Nevertheless, there was great hope that H. G. Wells' statement that the Great War had been 'The War to end Wars' was viable, and the sense of collective security and the League of Nations would resolve the war issue. In August 1928 the Kellogg-Briand Pact to renounce war was signed by Germany, France, America, and many other countries soon afterwards, but

---

[22] Howard, Michael, *The Invention*, p.55.

there was nothing in the pact as how to enforce its ideals. Within a few years of after the Pact, the Italian occupation of Abyssinia and the rise of Hitler were sowing serious doubts about the future, and the Japanese invaded Manchuria. In Britain the Labour Party had decided against all war but by 1937 had ceased its opposition to building up military forces. After the Second World War it was hoped, once again, that the Yalta Conference indicated that liberated peoples could create democratic rights of their own choice, and hopefully avoid the anarchy of states.

It was not to be with the postwar emergence of the Cold War with America seeing itself as keeping the peace and concerned about Stalin's version of communism. Britain and France had gone to war over Poland only to see it occupied by the Soviets and Germans, then by the Russians when Germany had been defeated. The threat of WMD created a sense of peace because nobody dared to fight, but proxy wars abounded. In Eastern communist Europe internal tensions continued, then wars in Korea (1950-53), Vietnam (1963-69) and when the Cold War ceased in 1990, it was promptly followed by the Gulf War (1990-91), the Balkan Wars in old Yugoslavia (1992-95), the war on Terrorism leading to the Iraq War (2003) and at the time of writing the Russian invasion of the Ukraine which again raised the threat of nuclear annihilation. During these last 100 years of wars there were countless other smaller conflicts across the globe, and recently yet more in Afghanistan, the Middle East, the Yemen, Syria, Algeria, Myanmar, Somali, Ethiopia, many of them civil wars, would be revolutions, and many based on pure aggression. It is only necessary to check a chronological global chart to see that since 1900 to the present day not a year has passed without a war in one country or another, many unknown about, and forgotten in the annals of global history. How many millions died in the Second World War remains speculation, vague figures between 50 to 60 million people and many more have died since 1945: again, an unknown figure. The only conclusion can be that nothing has changed in terms of war as it is part of humankind's character which is now reflected in most of the sovereign states.

It was anticipated that the threat of nuclear weapons would hold the peace which it appeared to do so on a global scale (with a sense of intense doubt during the Cuban crisis) and a new peace movement gathered force mainly under the banner of the CND (Campaign for Nuclear Disarmament founded 1957). After 1990 there were hopes the nuclear threat was anachronistic, but it was a forlorn hope with President Putin's attack on Ukraine. A new style of war known as Terrorism has developed, there are now three global superpowers following the decline of European power. China, America, and Russia are the powerful entities, and day by day there is not a country in the world which can claim to be free from the threat of internal or external war, not just because of the WMD, but back to the beginning of this study, because of the nature of humankind. It would seem from the above survey that those who claim war is inevitable and unstoppable may well be right, that war is part of our evolutionary nature. This sense of the inevitable begs the major question as to whether it can ever be brought under control, or if not finding ways to reach genuine agreements over the way war is exercised to limit the dangers. It would seem from history that international laws have failed time and time again, if only because people tend to look to their nation's leadership as the highest authority, so the question of why war cannot be avoided seems without an answer.

Have the causes of war changed?

In the modern era we have moved from the warrior leader to recognised countries, and we live in a global community of states as Rousseau once claimed. The states are like people, some good and some bad, but they must maintain relationships with neighbours which has been described as the balance of power, which still raises the question of why war starts if caused by a corrupt or bad state. Few people or states want war with its destruction, but as there was once the problem of belligerent warrior leaders, so now there are aggressive states, and the war of defence is inevitable. The efforts at international control have failed to avert war or resolve the issue, and the perplexities of this issue will be explored in Chapter Nine.

The Great War was not started by one leader seeking military glory it was an all-round bellicose explosion and the losing state Germany was penalised so harshly it led to serious consequences. The causes for war are limitless, complex and confusing, but the problems originate in humankind, within a state's make-up, or conflicting structures of various neighbouring states. Nothing changes under the sun, and humankind appears to have a biological propensity to fight. It is widely accepted that drugs and even drink can calm a man down to a peaceful state of mind, or alternatively wind him up to full aggression. It is almost an animal instinct and as in herds or groups of animals there is a pecking order based on strength or character, the age-old issue of the Alpha-Male. Psychologists have noted humankind's impulses to be competitive, driven by the need to dominate, hold power and be in charge.

States reflect their human inhabitants. History has recorded the development of states into an acceptance of politically defined areas who too often demanded expansion. This is not helped by the fact that there are elements in our nature which enjoys fighting, testing skills, and gambling for glory and prizes. For some it holds the excitement of a serious life and death sport where the opposition needs to be crushed. War is about who will blow the starting whistle, and the motives are manifold and varied, ranging from extending frontiers, increasing economic powers, political influence which is the new form of colonialism.

At the time of thinking about this chapter the United Nations Secretary General, Antonio Guterres in the Tenth Review Conference (2 August 2022) regarding the Non-Proliferation Treaty to prevent the spread on nuclear arms, warned the world had been lucky so far. He told his listeners that among the growing tensions around the world, no doubt hinting at Russia's attack on the Ukraine, China's demands on Taiwan, and North Korea's dalliance with missiles that humanity was just one misunderstanding and one miscalculation away from nuclear annihilation. It may be cynical, but even in this modern era the reality of human nature makes war an endemic humankind disease.

Nor have the causes for war changed. They still range from pure aggression, looking for power, influence, territory, and prestige, sometimes defensive or preventative. In the past and today they can be based on religious faith or an ideology, and often sheer fear of another state. Economics as a form of wealth has increased with major resources such as Middle Eastern oil and Russian gas, and the traditional pacts and alliances have not been helpful, as has been recently headlined with President Putin claiming NATO is encroaching on Russian interests and safety. Such pacts may deter war, but they can also start one. This can lead to dangerous diverse grouping, today with three superpowers and the various pacts, alliances, and influence they can bring to bear on other countries, too often leading to cross loyalties.

Many believe Wilfred Owen's poem *Dulce Et Decorum Est* stopped us making war romantic glory, it also identified war as evil, yet in so-called modern civilised times it still persistently hits our headlines.

Chapter Six - Can War Be Justified?
Jus ad Bellum

Brief introduction

There is a wide acceptance that any war of aggression for whatever reason is not only regarded as illegal by international law, but war is regarded by most as an unjustifiable and immoral act which brings destruction, death, and immense suffering. Wars of aggression as noted, are motivated by humankind's desire for wealth, territory, domination, political influence and often based on flimsy excuses such as it being necessary for the defence of the realm. Plato mentioned the possibility that a leader or government may start a war to distract from the misery at home. Some journalists suggested the Argentinian attack on the Falklands (1982) was based on this motive. Alternatively, they felt the Falkland Islands were their property (irredentist nationalism) as Spain may justifiably feel about Gibraltar, and the Italians about Corsica, but today this is understandably resolved by the wish of the inhabitants.[23] War may start because of an unjustified fear of a neighbouring state, even a sense that a country is somehow naturally superior to another. The reasons are simply manifold and nearly always dressed up with excuses such as defence, pre-emptive strikes, prevention, or intervention for non-existent humanitarian reasons. As in a domestic situation an uncontrollable belligerent neighbour may start a fight because he is a bully with the weaker neighbours, so in the global situation it may simply be a corruptly led country, sometimes called rogue states, and in a global community such neighbours are not only potentially dangerous but tend to exist.

There is little justification for starting a war, but it has long been regarded that just as an individual has a natural right to defend him or herself, so a

---

[23] Corsica once Italian became French when Genoa sold the island (1767) to the French who eventually incorporated into their state in 1789. Mussolini wanted it returned but his Foreign Minister and son-in-law Ciano noted in his diary that the Corsicans wanted to remain French.

state if attacked can justifiably go to war in defence of its existence. The nature of self-defence has been widely acceptable from the earliest of times as being justified, but even this situation has its limitations according to many thinkers. Human nature has managed to stretch the arguments relating to the boundaries of self-defence, sometimes with a sense of acceptable justification, other times raising serious questions. This chapter will explore the nature and limitations of self-defence, explore the argument as to whether a pre-emptive strike against the enemy can be justified or whether it is simply an excuse for a war of aggression. The debate regarding the nature of a pre-emptive strike has become more elastic as the war for prevention has been used on the grounds that the so-called enemy is preparing to attack. This may be based on mere political speculation, mutual animosity or even wrong information and intelligence. If a pre-emptive strike may be used as an excuse for a war of aggression it is much more likely with a war of prevention. Other areas of concern are wars of intervention which are often based on humanitarian grounds, but there must be clear international grounds of agreement as this can also be used for the wrong reasons. Finally, the vexed question of terrorism which comes under several guises must be noted, with some terrorists seeing their actions as defensive as they try and change unjust actions or abuse of their human rights elsewhere.

If humankind could ever manage to outlaw war and live in peace there would be no war crimes, no crimes against humanity, but history and humankind's behaviour appears to make this Utopia as unlikely as science fiction.

Self defence

It is generally accepted that international law may come into service in the postwar settlements of a conflict, but international bodies like the League of Nations, the United Nations, pacts such as the 1928 Kellogg-Briand, and international law courts have failed to stop war. The legal backdrop is subject to varying events and political changes can happen overnight. The only hope is some form of agreement based on humankind's sense of morality. This concept of a universal morality has been accepted from the earliest of times, made popular by Grotius who helped develop a natural law doctrine, which claimed people have inherent rights which are not endowed by man's law but by either God, nature, or reason. It is assumed that these individual rights also apply to states, and in Chapter Nine the question of a possible wide-reaching global consensus will be raised. It has been argued that moral rules can justify a state going to war, but it must always be in the cause of defence. In terms of justifying going to war the morality of the question is not adjudicated by soldiers but the opinions of mankind. Every leader who initiates a war wants to be understood as having a moral motive even though the decision may be painful. It sometimes takes an objective historical view based on hindsight to qualify if such actions had any justification.

From the earliest of times, as seen in the chapter on early thinkers, it is generally agreed that wars of pure aggression are unjustified, and that a more powerful nation should not attack and occupy a weaker neighbour. When in August 1990 Iraq invaded Kuwait, the world was generally horrified at this behaviour. Saddam Hussein was by nature an old-fashioned War Lord who had for geo-political reasons had had previous support from the West which undoubtedly gave him some confidence. The UN could only condemn and isolate Iraq which led to the vexed question of when war with retaliation could be morally justified. The current Russian attack on the Ukraine for most people remains highly questionable as it appears to be an act of aggression, but Putin, and in his time, Saddam Hussein appealed to the wrongs purportedly committed by the enemy.

Every country has the right to remain neutral and when it has proved to be so it should be honoured, which it was not when the Nazi German forces moved into Belgium in 1941. Not all approve of neutrality, especially if a neighbouring state is under vicious and unnecessary attack, which if based on imperial grounds should be stamped out if there is any chance of success. An act of aggression is always wrong, and it has long been a widely agreed understanding that self-defence against an armed attack is acceptable, and force can be authorised by the Security Council of the United Nations (Articles 39 and 42 of the UN Charter). Unfortunately, with the veto system and other reasons such international reaction is hardly enforceable in a major conflict. Defence is generally accepted as the only reason for war. A war of defence is the critical basis because people are assumed to have the right to fight for their rights and their lives. The territorial and political rights of a state are like the rights of the individual. When a state is attacked so are the rights of its citizens. Some resist by not fighting but by refusal to co-operate as once suggested by Bernhard Shaw, but more usually others will fight to the bitter end under such slogans as 'better dead than red'. Either way aggression is the outstanding moral crime and justifies defence using forceful resistance. Sometimes the power of the state cannot resist the aggression and thereby occupation, but resistance fighters and partisans may well have the moral right to fight back, as the rights of a state rest on the consent of the people, and not the occupying rulers.

The question of proportionality is an important factor worth considering. If, for argument's sake, nation 'X' has a border dispute based on irredentist nationalism (taking back lost territory) on a town which purportedly once belonged to the attacking state, then 'Y' the attacked state may justifiably defend themselves against the aggression. They may also have the rights in terms of a successful defence by 'Y' to demand some punishment or reparation from 'X'. It must, in moral terms be proportional, and the attacked state 'Y' would be unjustified if it continued the defence by crushing 'X' out of existence and having destroyed its military then enslaved the inhabitants. If armed soldiers had taken the disputed city, it can justifiably be taken back, but not by wiping out the attacker's population, most of whom would be totally innocent. Or to use a somewhat trite example, using a battlefield nuclear weapon against angry tribesmen, or more realistically as Mussolini used gas against Ethiopians in 1936, and Hussein against the Kurds.

The emergence of terrorism in the modern era has often led to asymmetric warfare, where a small and impoverished nation, presumed to have been the home of a terrorist group can find itself attacked by a major nuclear power. This can lead to the moral question of proportionality, and the invasion of Iraq following the 9/11 terrorist attack on the Twin Towers is viewed by many as falling into this unprincipled category not only because of proportionality, but for other moral reasons which will be raised later. Proportionality is a difficult issue not only in terms of self-defence as many factors come into play, but also matters which will be raised in Chapter Seven on the conduct during a war.

Pre-emptive war

There are distinctive dangers in whether a pre-emptive strike is morally justified. The main dangers facing whether this type of war often stem from bellicose politicians making threats, a feature of the past and ongoing to this day. Such rantings do not necessarily mean military action will follow, nor the neighbour's building up of arms mean the same things, as this may be pure defensive measures. There may have been previous agreements not to build up forces, but doing so, and even mild blockades do not necessarily necessitate war. Before planning a pre-emptive strike, it must, to be morally justified, be based on defensive grounds which would demand there are hostile intentions, and an attack is deemed by all intelligence and opinion to be essential. This is one of the reasons that the current tensions surrounding North Korea are always so high and causing anxiety. One may have a noisy rude neighbour always shouting insults and making threats, but it must be clear that it is only noise, especially if he tends to stay his side of the boundary. The best defence is to build up one's own defences.

In World War II Norway's long Atlantic coastline and its natural resources worried Churchill and he was preparing to mine or even take over their ports, which would have been a pre-emptive strike both against Norway and possible German expansion, but the Germans struck first. When King Haakon VII of Norway was told they were being invaded he asked if it were the Germans or the British. The moral implications are confusing and could be debated either way given the circumstances of the day. In the later Nuremberg Trials, the German defence argued that it was a pre-emptive strike because they knew of the British plans. In history there have been many pre-emptive strikes, some with justification, and some without that qualification, and most of them dubious. The expression of a sliding scale is often used to underline this problem.

In June 1967 it was clear that Egypt was going to blockade Israeli shipping and evidence was accruing that there was to be an attack on their country, so Israel launched a pre-emptive devastating airstrike which resulted in the Six Day War (5-10 June 1967); there were many casualties, but the pre-emptive strike may have saved a longer war with more casualties.

Pre-emptive warfare usually begins because of international turmoil, and the term is often abused. In what was sometimes called the Manchurian or Mukden incident the Japanese claimed to be fighting a pre-emptive war against the Chinese, who, the Japanese claimed had blown up the South Manchurian Railway at Mukden and they were merely defending themselves, with later evidence suggesting the Japanese had blown the railway up on purpose. This problem also emerged when Nazi Germany claimed Poland attacked first when they had used the SS and prisoners in Polish uniforms to attack a border post, known as the Gleiwitz incident. It was a matter of abusing the pre-emptive defence arguments to start a war of pure aggression.

It is generally accepted that for a pre-emptive strike to claim a moral basis the magnitude of the threat needs scrutiny, the need to look to alternative means instead of force, and a close examination of likelihood that the threat ought to be regarded as certain, and it must come under the justified claim of self-defence. Most states would prefer to err on the side of caution but if the threat is obvious to everyone, then reaction should be prompt to avoid further risk, and be able to claim moral justification. The trouble is the term pre-emptive self-defence should now be regarded 'as a tree with a number of branches. These branches are as follows: anticipatory self-defence, preventative self-defence, and precautionary self-defence', all of which are suspect and open to easy abuse.[24]

War of prevention

The nature of preventative war has much to do with the concept of the balance of power often coupled with a sense of fear. It is not a chemical or mathematical equation as political or diplomatic reasonings can be deeply flawed, never precise and potentially disastrous. The sense of fear

---

[24] Kinsella D, Carr C. L. (Eds) *The Morality of War, A Reader* (London: Lynne Rienner, 2007) p.144.

may be a complete misunderstanding or a serious misreading of the situation. The balance of power, which is to do with a sense of a reasonable distribution of power and assets can be equally misread. At the personal domestic level, it is like watching the Jones next door having concrete delivered and wondering whether it is for a pill box, launch pad, or a swimming pool. The balance of power can be reassuring as in times past it has been believed to hold the peace, but it can never rid states of the fear of their neighbour's intentions. It raised the question as to whether the guidelines of an imminent threat can be moved or stretched to the mere possibility of an attack, a purely theoretical confrontation, but even in an anticipatory concern there may be lurking dangers. North Korea's experimentation of firing a long-range missile across Japan (October 2022) naturally made neighbours nervous if not antagonistic. It is often an impossible task to understand the motives of other people and states. When Hitler amassed troops along the Soviet frontier most people (except Stalin) realised the German dictator was about to start Operation Barbarossa, and many believed and hoped that when the same scenario recently occurred on the Russian-Ukraine border, that Putin's claim it was a military exercise was honest and could not happen. The danger of speculation is that it cannot always rely on knowledge and Intelligence. It has often been suggested that before any resort to war that sanctions should be applied first, and many pacifists have argued that before the Gulf War (Desert Storm) insignificant time was given to this policy, as sanctions could be described as a non-lethal indication and signal of the global political will.

Nevertheless, it has been argued that a country has the moral right to make a war of prevention, but it is very different from a pre-emptive war although some feel it is an overlap or mild extension. In the modern era it has emerged over the question of terrorism often motivated by religion ideologies or protesting at the occupation of their homelands. This has meant attacking major powers such as the twin-tower disaster, known as the 9/11 (2001) disaster which left America reeling and uncertain as to whom the enemy were, or which country had harboured them. In September 2002 in an American report by The National Security Council it stated 'new deadly challenges have emerged from rogue states and terrorists', which it qualified as brutalising their own people, having no regard for international law, acquiring WMD.[25] The report continued by stating that 'it has taken almost a decade for us to comprehend the true nature of this new threat' and 'the United States can no longer rely on a reactive posture'.[26]

As with a pre-emptive war it must be clear that military action would stop the danger, and that the location or country is undeniably right. In terms of terrorism this is almost an impossible task as they can move location or country with considerable ease, and whether a state can be described as rogue raises further questions as to bias, misinformation, or lack of understanding. It is therefore not surprising that the war of prevention has yet to be recognised in international law, clearly begging the question as to whether a war of prevention can be considered as justified. A genuine pre-emptive war starts because it has been established the threat is impending beyond doubt, but in a war of prevention it is to do with merely a possible danger or threat, which may or may not be realistic. Is the existence of a possible threat justifiable in moral terms is a vexed question. When North Korea fires its missiles, it must be an unnerving experience for neighbours or is it the leader boasting to his people and telling the world he is important or is the threat real; a question where it would be wrong to 'jump the gun'. Before a military reaction there must be a certainty that the threat is real, but unlike pre-emptive war it would be more difficult to react with the necessary certainty. The theory of a war of prevention lacks the certitude relating to pre-emptive wars.

In terms of terrorism there must be a clear and indisputable connection between the terrorist and the so-called host country. Because it is generally understood terrorists would like to have a WMD, any country capable of producing WMD or the technology and science could be put under the microscope, even though they may be innocent, especially if they have some form of instability.

The debate still rages around the decision to attack Iraq following the 9/11 disaster, with President George Bush telling the UN General Assembly that Iraq was a grave and gathering danger, later proposing an attack on Iraq as

---

[25] See Kinsella, p.127.

[26] See Kinsella, p.129.

a form of anticipatory self-defence, but better expressed as a possible emerging threat. He was basically claiming that the traditional guidelines on making war were behind the times with the emergence of serious terrorism, even though his final decision clashed with the UN Charter, prompting Kofi Annan the Secretary General of the UN to state it was illegal (16 September 2004). The Americans, supported by the British, suggested that the UN had authorised possible military force over weapon inspections, argued that the West had an inherent right to defend itself, and finally the reason of relieving the Iraqi people from a dangerously repressive regime under Saddam Hussein. Despite objections from the UN the war went ahead, and by doing so questioned the established apparatus of the UN, causing many to see the international structure as a mere talking shop with no real authority.

The Bush administration with the support of the British government under Prime Minister Tony Blair, won a military success, but it was almost an asymmetric war with a desert country fighting against the vast technology and numbers of a Western power. The promised WMD were never found apart from degenerating chemical weapons, there were some 4,000 plus American casualties, and among the Iraqi population an estimation of over 400,000 killed. This death toll and the money involved raised the old question of proportionality. The danger was that this form of war could be regarded as setting a norm which many found unacceptable, for others a possible excuse. It was trying to set a norm which was somewhat equivocal and therefore in legal terms difficult to establish. It meant that the USA and Britain had set a precedent for other states, who could react in the same way, claiming an anticipatory attack was underway. It also managed to increase tensions on the international scale giving scope in such fraught areas as India clashing with Pakistan, and even China looking to Taiwan which today is becoming more serious.

At the time of controversy of the war on Iraq (2003) many felt this type of war unjustified, and it was aimed at an innocent party in terms of its suggested connections with al Qaeda. The writers John Mearsheimer and Stephen Walt wrote a convincing article that Hussein was not the threat that Bush and Blair painted.[27] They explained his war against Iran (1980)

---

[27] See Kinsella, pp 148-154.

was more defensive based on the religious aggression shown by the Ruhollah Khomeini, even finding a motive for his invading Kuwait, and that Saddam could be deterred or contained. During the Gulf War Hussein had ordered Scud missiles be aimed at Israel and Saudi Arabia but not armed with WMD. He had used chemical weapons at the resident Kurds, but it was argued that he was unlikely to use them against a superpower. Bush's claim that Saddam would use nuclear weapons to blackmail others was unlikely not just because he did not have any, but he knew that it would be too dangerous. Many knew that there was no possible connection with Osama bin Laden who held a deep-rooted hatred for such desert rulers as Saddam Hussein, who was a despot but one who would not risk a war against a superpower. This so-called war of prevention was not only questionable but was based on a trumped-up analysis of the issues focused on Saddam's despotic nature.

There were serious protests on a global scale, and in America and Britain there were major demonstrations against the action. For many Americans President Bush's reputation is smeared by his action, and in Britain it was followed by a seven-year Chilcot enquiry, which stated Iraq had not posed a national threat to Britain, that the claim he had WMD was unwarranted in its certainty, with America and the UK undermining the authority of the UN. Tony Blair's decision has hounded him since marking him as a failure when he may have been viewed as one of Britain's better Prime Ministers had he not backed the President Bush administration. Either way it made a war of prevention not only politically and legally suspect, but in the cost of lives and failure to stop terrorism it cast huge doubts on the morality or acceptability of Prevention War.

This war was based on the terrifying attack on America with such ramifications it created a sense of anger and fear. The nature of human fear about a neighbour has often caused war, but the fear has increased when there is uncertainty about as to who the attackers were and from where they came. Such was this fear following the attack on America it started to change the guidelines of a Just War. The theory of Prevention has blurred the picture of self-defence in giving a pre-emptive war more strands than an obvious impending attack.

This problem of tackling terrorists hidden around the world in countries which might or might not know of their whereabouts is a vast void for dangerous speculation, especially since reliable intelligence sources are probably short of appropriate evidence. It led to the occupational war in Afghanistan war which started in 2001 and finished abruptly in 2021. The Taliban were toppled by various means, not least by supporting their internal opposition of the Northern Alliance which was also probably guilty of war crimes. The irony was that as soon as President Joe Biden suddenly extracted Western troops the Taliban came back to power instantly. The so-called war of prevention as enacted by President Bush has started a change in the guidelines for a Just War, challenged what authority the UN had, and made an unstable world more fearful. It was a failure in so far that it failed to eliminate terrorism which can grow like a weed in hidden crevices which even the householder cannot see. Terrorists believe in their cause, be it ideology or faith, and the inequality they may have witnessed, and only some form of universal justice which is sadly lacking, will heal such problems. It is known that some terrorist groups want to find WMD such is their hatred, realistically it demands action, but never at the expense of the innocent. Adequate surveillance and genuine international cooperation based on a sense of the community of nations might possibly achieve such an end without too much bloodshed.

Intervention and humanitarian intervention

Armed Intervention in the affairs of another state to change its domestic situation has a long history of political motives, often concealing an immoral war of aggression. However, one which may have a possible moral reason is when armed intervention occurs to stop widespread unnecessary suffering. Nevertheless, even this principle has long been challenged, namely that countries should not intervene in domestic affairs of other states. In his book *Liberty* (1859) John Stuart Mill proposed that states are self-determining communities whether they are internally free or not, and it is for the members of that state to seek their own freedom, as they had the right to liberate themselves. Later he may have had a change of mind (circa 1874) when he considered intervention to help people struggling under a tyranny, especially if it were a foreign domination. A hundred years later the issue remained fraught, and in 1970 the UN stated that 'no state or group of states has the right to intervene, directly or indirectly, for any reason whatsoever, in the internal or external affairs of any state'.[28] Imagine in a state where there is a vicious despotic leader whose god is Hitler and that leader has decided to murder on an industrial scale all Jews, foreigners who cannot speak their language, and people of a particular skin-colour. Such is the heinous behaviour of the situation many people with strong moral leanings would want a UN armed force to intervene. This is a persuasive sentiment of saving innocent lives which is based on a sense of universal morality, demanding some form of humanitarian intervention. In some circumstances it could still be dangerous as in the past when Lord Palmerston as prime minister thought such intervention would shatter the peace of Europe. Had NATO interfered when eastern European countries wanting freedom from Soviet domination it might have provoked a war of annihilation. The situation persists to this day with the Russian attack on Ukraine when direct intervention might lead to nuclear war. It also has left the possibility of UN intervention in a complex situation, especially because that UN intervention is probably the only way forward. The Security Council during the Cold War with its veto and sensitivies to global safety left the UN tied by their hands and feet. In addition to this the UN Charter (Chapter VII) severely limits if not forbids the UN role in a state's domestic life. This was apparent when Saddam used gas against his resident Kurds (1991), but then the Bosnian conflict (1992) when NATO felt obliged to step in, and the Rwanda civil war (1994) which was so evidently inhumane that the UN permitted some states to operate 'relief operations', France in Rwanda, and America in Somalia.

This raises the bogeyman of civil war when it is generally accepted that outside states should not intervene in a society at conflict with itself. Even this is not a black and white situation, as Britain and France stayed politically outside the Spanish Civil War (1936-39) which allowed Germany and Italy to turn the balance in favour of a totalitarian despot. The motives for interfering in a civil war are often highly dubious and too often political. When America entered the Vietnam War (1955-1975) it was not for humanitarian reasons, but its fear of communism extending its influence,

---

[28] Kinsella, p.187.

and they propped up a corrupt government against their fear of the communist option.

If some neighbours in a street start to fight, other people can try to stop the mayhem, but it can sometimes escalate out of proportion and lead to more damage. The only authority which can control the problem is the state authorised police service (or force) which in global terms is the UN uncertain of its own rights. It is also difficult to identify when it is a genuine civil war in which the citizens are working through their problems, and as to when their behaviour becomes inhumane. Each state has its own characteristic background and norms of behaviour.

For intervention to have any moral basis it must be based on purely humanitarian motives, but that raises the question of defining inhumane behaviour. There are some violations which for most people are obviously totally abhorrent and inhumane such as the industrialised annihilation of a race of people as experienced in the Holocaust. This is based on that sense of a common morality, sometimes covered by the term natural law (see Chapter Nine) which relates to the whole of humankind and not just a state or one community. It is natural law which justifies self-defence, and condemns slavery, mass-murder, unjustified killing, and, it is proposed, should mean coming to the aid of others in distress. On a personal domestic level if a passer-by sees a helpless child being drowned in a pond by an apparent aggressor most humans would interfere to save that life. The law of their country may state that such action is commendable or even a duty, another nation may stipulate that interference could be undue meddling. This raises two major issues, the first is that for most of humankind saving a helpless child is paramount, the second is if the law is itself wrong then there exists an inhumane state. Natural law with its sense of a common morality, however obscure for some to define, can be the only answer.

Correcting a wrong has already been noted by previous thinkers with Aquinas noting that war must be a means of correcting wrongs, and this theme has been reflected down the ages, offering the issue of extending a Just War to a just intention, and the old claim that it must be initiated by the sovereign, leaving the UN today the only sovereign with such a power. The main issue is that the UN, as noted above, cannot react like an ancient sovereign, and is bound by rules which from many points of view leaves it hamstrung.

Nevertheless, it is within the instinct of humankind to help others in serious danger if is done within moral bounds. In the domestic situation of the child being drowned in the pond by some aggressor, it must be ascertained that the aggressor is not ducking the child for mutual fun, and that reason is applied in the first place by persuasive talk. It is also critical that only proportionate force is used to stop the death, not necessarily by shooting the attacker, but by restraint and hinderance because killing may escalate a wider situation, or the attacker may have suffered a psychological illness; killing must be the very last resort when all other means have been exhausted. At a personal level it is always confusing, at a national level potentially dangerous and complex, but the instinct of a common morality is starkly obvious.

When the Israeli government authorised the armed raid on Entebbe airport (July 1976) in Idi Armin's Uganda it was to save lives and for no other motive, and most people found this acceptable. A similar example occurred in the invasion of East Pakistan (1971-Bangladesh) when the Bengalis were suffering and seeking assistance, the Indian army defeated the Pakistani force and promptly left not making any political changes or any other form of interference. The intervention was for humanitarian principles only, as the Pakistani military and Islamic militias (Razakars) had killed unknown thousands of people and carried out what has been described as a genocidal rape.

For intervention to be described as moral or justified at state level it must be something which really disturbs humankind's conscience. Many will be unhappy with the way some states treat women, or other states imprison political activists, or execute people for minor infringements and homosexuality, or deal with peaceful demonstrations, but for armed intervention it must be based on a gross humanitarian crime such as genocide or ethnic cleansing, otherwise it might provide some states to use war for unjustified reasons. Such intervention should not have political or economic reasons, but be based on pure humanitarian motives, always seeking stability and peace. There should be no motive of changing the balance of power, or the political balance within the state. It should always be treated as the last resort and utilised only if success is likely. As in other areas of armed conflict it should be proportional and there should not be any abuse of commonly held beliefs, and morals must always be proved to be justified by having the right intention. Armed intervention should only occur when all other avenues of resolving the problem have been exhausted, and should be based on a collective basis, not committed to a single nation.

The difficult question is who should intervene. It seems to most people that as the problem is dealing with the mass human conscience and the sense of a common morality that it demands an international collective response from the only body authorised by agreement to remedy this issue, namely the United Nations. The inability or failure of the UN to offer a way forward led to the NATO intervention in Kosovo, and on 13 October 1998 NATO issued activation orders for the execution of limited air strikes and a phased air campaign all of which was made outside the UN structure. This raises the nature of the force deployed which must be proportional and follow the other accepted rules of war, nor does it mean that once action is taken that the forces stay with the motive it might happen again, but like the Indian army in East Pakistan mentioned above resolve the problem and instantly move out.

It is not helped that there are no agreed rules on this issue either when intervention is needed nor how the situation ought to be handled. Given humankind's propensity for extreme immoral behaviour this feels like a serious failure, and complex arguments still rage to this day, when all that is required is establishing the right motives for intervention and the normal responsibility to protect the helpless. It has been suggested by several of those concerned that the word intervention should always be preceded by 'humanitarian' to make the motives acceptable. The issue for the UN is their agreement on the total rights of a sovereign state (UN Charter Article 2:7) but some states appear to neglect the foundation of that sovereignty which includes caring for its citizens. It is simply a matter of the UN establishing some basic principles of when humanitarian intervention may be deemed internationally legal.

This problem of intervention raises the issue that failure to establish a known and realistic policy means that when nations forge ahead on their own initiative (NATO in Kosovo and President Bush in Iraq) it undermines the authority and usefulness of the one established international structure of the UN. There have been many times when intervention has happened in the past thirty years, and unless the UN can face its own structural problems and arrive at an internationally acceptable policy then humanitarian intervention instead of being a game-saver for humankind could too easily lead to disaster.

Terrorism

Terrorism is and will remain a highly contentious area, but it is necessary to raise questions which for many people will be both challenging and anger-making. For most people news of a terrorist attack appears in the headlines for a time and is quickly forgotten. When the 9/11 attack occurred in America the number of deaths and destruction had a marked effect with a major atrocity on the doorstep, not only in America but globally, and terrorism has remained a major issue since. The very word terrorist has a pejorative undertone and is rejected as totally immoral by most people.

Terrorism has been defined as the unlawful use of violence and intimidation, especially against civilians to pursue political, ideological, or religious ends, creating fear to terrify people into submission. However, there are many definitions, and points of view. In his book on *An Anatomy of Terror*, Andrew Sinclair gave terrorism a wide range of activity and from the earliest of times within recorded history.[29] He rightly conveys that terrorism can take many forms ranging from criminal terrorism (Mafia and so forth) to all those mentioned above, and including resistance and partisans fighting an occupation, to terrorists utilised by governments, to one-off actors representing small groups with ideological or religious views.

However, in terms of the modern era, it has four characteristics according to Alex Bellamy, first it is politically motivated violence, secondly

---

[29] Sinclair Andrew, *An Anatomy of Terror, A History of Terrorism* (London: Pan Books, 2004)

conducted by non-state actors, thirdly it intentionally targets non-combatants, and finally achieves its aims by creating fear within societies.[30] This is a survey open to challenge, as religion can be a major motivator and the most senseless, the target is not always non-combatants, but these are minor comments and Bellamy's four classes are generally sound.

When on 10 October 2022 President Putin ordered the missile attacks on Ukraine's capital and other cities, he explained it was in retaliation for the 'terrorist attack' on the bridge connecting Russia to the Crimea, thereby leaning on its pejorative sentiments for support of his actions. By using the word terrorism, he was suggesting some form of justification because terrorists today are almost seen as the universal enemy. Terrorism has an ancient global coverage, but in this study the main interest will be those acts of terrorism in the modern era to this day and raises yet more issues in the pursuit of justified war. Because of the varied nature of terrorism, it is not always easy to define. It is often stated that terrorists aim at civilians or non-combatants, but it must be noted, not always. In October 2000, a guided missile destroyer the *USS Cole* suffered a suicide attack while refuelling in Yemen's Aden Harbour. In October 1983 over 200 were killed in the American marine barracks in Lebanon, and even during the attack on the twin towers in 2001 one plane was crashed into the military defence centre of the Pentagon, and during the Irish crisis the IRA targeted combatants and non-combatants. Whether civilian or military the intention of terrorism is to create fear. Terrorism is considered immoral because it attacks civilians, but it must be remembered that more people have died from government authorised actions. In making this distinction between innocent citizens and military personnel there appears to be a sense of hypocrisy as states have done the same thing, not that it is therefore justified.

Terrorism is condemned for using fear, but widespread fear is present when a state indicates it is going to make war on its neighbour, and the assumption made by the majority that terrorism is totally immoral does not always hold water when compared to some states. When the Russian bridge to the Crimea

---

[30] Bellamy, Alex J, Just Wars (Cambridge: Polity Press, 2006) p.136.

was blown up by the so-called terrorists it was a tactical annoyance to the Russians, but the retaliatory missile strikes created more innocent deaths and fear amongst civilians. Had a terrorist group managed to kill Idi Amin it would not necessarily be regarded as unjustified any more than the 20 July plot against Hitler. Others believe that terrorism's use of violence never produces results but only makes matters worse. This is a criticism which could be aimed at state governments, both in their domestic lives and in their resort to war, the situation in Afghanistan has not improved, the crushing of Iraq killed too many innocent lives and resolved nothing, and the Russian attack on Ukraine has produced untold deaths and turned a once respected state into one of pariah status.

The writer Alex Bellamy cogently argues that there are 'three distinct moral types of terrorism'.[31] First, it is possible to have a sympathy for those called terrorists when they are working to correct a manifest evil within their own community. Some states are not only dysfunctional but seriously defective and creating a serious breach of human rights within their own community. The classical example of more recent times was the gross apartheid system in South Africa which was not only immoral but created fear and suffering. Nelson Mandela who became the first President of South Africa (1994-1999) spent years in prison for seditious work against the apartheid government and was convicted as a terrorist, which Bellamy describes as partial terrorism. Secondly, in what Bellamy describes as a 'grey area' is typified by the Palestinian-Israeli conflict, and finally Al-Qaeda which he rightly addresses as pure terrorism.

Amongst those who study terrorism are some who raise such questions as the consequences of this form of aggression. If for example a terrorist is working to resolve an unjust situation in his homeland where there is considerable suffering, and his work helps alleviate the situation his work could be seen as justified for some. It does not alleviate the issue that in accomplishing the task that totally innocent people suffered, but on the other hand a war of intervention supposedly based on correcting a wrong innocent people also die. These comparative judgements are easily

---

[31] Bellamy, p.136.

conjectured from the comfort of a study, but they must be given the light of day however indigestible they may be.

Again, the consequential type of argument is for most irrelevant, but it tries to draw attention to a violation of human rights, as many may argue is the case of those suffering in the Palestinian context. Whether it is justified to use extreme violence will not go away as many will argue that there are other means to resolve such issues apart from resorting to violence and fear. The terrorist may argue that it has never worked, and the blind eye of the world needs to be opened to the injustice. The lack of human rights and the unfair distribution of power may draw some sympathy, but the issue is always besmirched by using force against innocent third parties.

Another terrorist classification is that where an ideology or religious faith is the driving force, which, understandably for most, is a highly objectional area. The one which has constantly been in recent focus is that of Al-Qaeda which is a militant Sunni fundamentalist form of Islamic extremism who view the Western world as an alliance of Christian-Jewish oppression. This is in this writer's mind unquestionably the most intolerable form of terrorism having lived in a democratic city where Church spires, Synagogues and Mosques lived with their communities peaceably alongside one another, even having 'open-days' so one faith could see how the other worshipped. Tolerance of another's religious faith would cure many problems. The argument that the leaders of some powerful nations are of Christian or Jewish background does not mean that it is their faith which is the driving force, but rather their politics.

The nature of terrorism demands the use of violence and according to the writer Igor Primoratz 'it targets two different persons or groups of people. One is the primary, the other the secondary target. The latter target is directly hit, but the main aim is to get at the former, to intimidate them into doing something they otherwise would not'.[32] In other words a terrorist may hold a group of innocent bystanders as hostages, but it is to draw the attention of the world and specific political leaders to wrongs in the terrorist's home area.

---

[32] Kinsella, p.169.

Nevertheless, it is argued that even in the distant origins of the Just War theory non-combatants should not be killed, and the killing of remote third parties nearly everyone finds repugnant. A hostage on a hijacked aircraft might well explain to the gun wielding terrorist that he knew nothing about the terrorist's home injustices, never knew anything about the country, and why does he not go home and stir up a justified rebellion there? It is a meaningless argument to the typical terrorist who is using a second country to bear influence on his home area. A terrorist cannot allow him or herself to show any respect to those under threat, nor pity. It is this which defines the nature of a terrorist. It is usually the major powers, especially the West who are the targets because of the influence they carry, it would be senseless for an act of terrorism in places like Tonga or Fiji unless it were an internal matter concerning Pacific islands.

It has been held by some that terrorism can be justified under some circumstances but raises the question as to whether to create justice in one sphere is it right to perpetrate injustice elsewhere, especially when it violates the rights of innocent third parties. It may be more justified if the terrorist's target were the area and people where the injustice is happening, based on the belief that it is immoral to use force against the innocent. It is the age-old argument as to whether the end justifies the means.

In his book Alex Bellamy concludes that terrorism is never justified, and most people would agree with his statement.[33] However, we are faced with grey areas as always. It is impossible not to have some feelings of sympathy for those non-state terrorists who are fighting against a major injustice such as apartheid in the old South Africa, but terrorism has taken on a much more brutal form. While there is widespread sympathy for the Palestinian situation the attack at the 1972 Olympics in Munich when Black September took nine members of the Israeli team, killing two and naming the operation Iqrit and Biram, (two Palestinian Christian villages where residents were expelled by the Israel Defence Forces) it invoked no sympathy for their cause. The killing of young athletes caused deep resentment because all attacks where third parties who are totally innocent

---

[33] Bellamy, p.156.

are killed to create a sense of fear in the hope problems at home are resolved or revenged are despised by most.

There is a degree of hypocrisy in relation to non-state terrorism. All states condemn terrorism because of the sense of blackmail, fear, and killing non-combatants, yet most states have committed these various crimes. There are many examples, perhaps the most singular and obvious examples were the strategic or carpet bombings of Nazi Germany which induced fear in the population with the intention of breaking their willpower. Many have argued that bombing has become more accurate but still innocent people have died. The bombing in Vietnam was to induce fear and killed many, as do the Russian missile attacks on Ukraine. This does not give the terrorist justification because such bombing may be regarded as unjustified.

For those who argue that the consequences are a critical feature they rarely work with a few exceptions. The 9/11 attack was in many ways both wrong and irrational and resulted in a typically misguided human response. In America nearly 3,000 died in the attacks, but Iraq was wrongly blamed and in the region of 400,000 died in a country which Osman bin Laden despised.

Some have argued that perhaps terrorism cannot necessarily be ruled out as unjustified given the appalling conditions and injustices to be found around the world, but in terms of the common morality killing innocent third party bystanders is wrong and can never be justified.

Chapter Seven - Rules of Battle?
Jus in Bello

Introduction

If the moral and legal guidelines for going to war (*Jus ad Bellum*) remain controversial, the so-called rules for conduct in war (*Jus in Bello*) are traditionally better defined. In the distant past there were rules of chivalry, but they were not classified laws and were at the whim or nature of the military princes of the day. As humankind's history unfolded with weapon technology advancing, wars created more casualties and deaths. The rise of nationalism and increase in manpower meant the glamorous knights of the past left the stage, the brutality increased. Chivalry is an aged-old concept of knights fighting knights, but on the battlefield with the foot-soldiers it was, as Shakespeare's Falstaff noted, a time of death and injury. From time-to-time elements of chivalry still emerged even in the modern era, but they were limited as the nature of war changed, especially with killing from a distance and not seeing the enemy face to face. Soldiers who can see their opponents in the trenches and occasionally fight hand to hand see the enemy as another human being which emerges in Wilfred Owen's poetry. Michael Waltzer points out that this age-old sense of chivalry persisted with fighter pilots, and there were elements of this human feature in the unofficial Christmas Day truce in the 1914 trenches.[34] The modern era has increased the distance between combatants, with tanks, missiles, bombers, WMD, and the chances of chivalry are reduced as the enemy becomes an object rather than a fellow human-being. In the use of drones the enemy becomes an object on the screen like a child's war game on his iPad or computer screen.

It was deemed necessary from the earliest of times to try and produce rules of conduct for those fighting, both at command level and the combatants in the battlefield. These are usually epitomised by the later formulated conventions of the well-known Hague and Geneva Conventions with their subsequent protocols signed by many nations but not all. Their intention

---

[34] Waltzer, Michael, *Just and Unjust Wars*, 5th Edition, (New York: Basic Books, 2015)

was to be civilised even in an uncivilised situation, and such are the fraught and deadly machinations of combat which create anger and hatred such laws are often broken. Combat in the field is primitive even with modern weapons, split second judgements must be made, fear, revenge, and hatred dominate, and the call for victory at any cost remains prominent, generally known as 'heat of battle' crimes, so it is perhaps surprising that rules drawn up by even the best minds in the sanctuary of peacetime chambers are often neglected on the battlefield, not least recalling that soldiers are in a very different context to the domestic criminal, and placed in their position by their country's leaders. These codes are often referred to as war conventions, and a failure to observe them has led to many national and international courts, but there have been times when even at the highest level of command decisions have been made to break with convention to avoid defeat or claim victory. This type of decision made by senior command tends to come under greater scrutiny postwar than with ordinary combatants making bad decisions in the heat of the battle.

At an overall level it is agreed that soldiers have equal rights to kill the enemy, but to avoid butchery for the sake of revenge, the argument for proportionality emerges, the right to surrender should be always accepted, that POWs (Prisoners of War) be treated fairly, that combatants need a uniform or insignia to identify themselves as legitimate combatants, and the demands that non-combatants and human rights should be respected. The list is extensive but as on the domestic scene as criminal law is frequently broken, so in the heat of battle and the desperation to win at any cost means the war conventions are frequently ignored.

Human nature is more diverse in character than any other form of life, and the good, indifferent, and bad permeate any community and its military. It is well-known that prisons holding life-threatening criminals have been a resource for the military, and some men volunteer for service who have a proclivity for excitement and violence, and others who have little understanding or sensibility about a moral conscience. On the other hand, many more soldiers are often reluctant to kill, especially when they see the enemy is just like them, as often implied by many of the Great War poets.

Francis (Franz) Lieber (1798-1872) a German American jurist who witnessed the American Civil produced the Lieber Code for governing armies in the field. His work has been regarded as the forerunner for the Geneva Conventions, and he demanded that guards at an outpost should not be shot but a gun be fired to drive them inside, which intimated that sniping and even ambush was morally wrong. The American Civil War has often been regarded as the first modern war, but since then wars have become more deadly, destructive, and facing possible annihilation. Human nature has not changed, and the moral issue remains a dilemma.

The POW rules

As the years have passed the conventions of war regarding the treatment of POWs has vastly improved. The Third Geneva Convention (12 August 1949) reinforced that prisoners are treated humanely (Article 13), have respect for their honour (Article 14), not be asked for any information beyond their personal details (Article 17) and so forth. From the start of the earliest war conventions these rules have often been abused even when the states were signatories to the various conventions. Without these conventions the barbaric treatment even in modern times almost defies belief. As recently (in historical terms) as WWII Nazi Germany starved and worked Russian POWS to death, and the Russians still held over two million German prisoners until 1950, releasing the rest in 1956, more than ten years after the war. Japanese brutality in their camps has been well documented, and major questions have been asked about the conditions of Allied holding camps of German prisoners during the collapse of Nazi Germany. Despite the efforts of the international community being a prisoner in Vietnam, by either side, was far from humane, especially when they became bartering fodder. Much has been made of life in prison camps in autobiographies and films, and it seems clear that the individual commander of these camps played a major role. In some it was evident that the level of humane action or brutal behaviour was dependent on the leading individual which raises the question of whether it is possible to set a morality bar, which is unlikely as it is to outlaw a criminal activity, so a police force is not required. At least setting a convention makes it clear when it becomes illegal by the state, or the individual can be indicted which may act as restraint in another war. The rules were broken time and time again, escaping prisoners were shot, some executed, and Hitler gave the *Commando Order* that such men should be executed even in uniform. It also carried another implication, that if one side ill-treats its prisoners the other side may respond with similar action, which is the old theory that we will not use gas unless you do, making reciprocity a two-edged sword. After the Dieppe Raid (August 1942) questions were raised about the treatment of prisoners, British Commandos had tied up German prisoners, and it has been suggested that was why Hitler gave the infamous Commando Order (October 1942) for Allied commandos to be executed.

Treating a prisoner as lawful or not remains a major issue to this day. When fighting occurred in Afghanistan the Taliban never wore traditional uniforms or Insignia and the Bush administration therefore considered them illegal, and because of the lack of black and white clarity it led to the infamous Guantánamo Bay prison in Cuba. It was argued they were treated humanely, but serious doubts have been raised on this issue, especially over the use of torture. The Bush Administration 'made it clear early on in the war on terrorism that the conflict with Al-Qaeda, in Afghanistan and elsewhere, falls outside the confines of Geneva Law and, by implication, the theory of the just war from which it derives'.[35] This was hardly a moral standpoint as the administration was saying we shall act in a moral fashion only if you keep to the same standards. This implied that America could apply torture to which America had signed an international convention against this means for extracting information.

Necessity of torture

The argument is that torture can save lives, and if it means millions of lives can be saved by extracting the necessary information from one prisoner then many might agree, albeit with some embarrassment or conscience.[36] If a million lives are at stake and one man can stop this, then the difficult question is raised whether a hard-line moral stance is right, or whether pragmatism is needed. This dilemma often referred to as the ticking bomb raises very serious questions within the realms of law and morality. The question of the use of torture raises the issue that immoral means are deployed for the overall good of humankind, and it has become sharply focused on the question of a WMD given the scenario above where one prisoner has the means to stop the killing of millions if not the destruction of a country. Torture has a sordid past in every nation, even the Church in the days of the Inquisition used it to save souls, but today there are both domestic legal restraints and treaties which outlaw torture, yet it is known that many countries use this means, some openly, others behind closed doors.

In an essay on this subject Alan Dershowitz pointed out that before the 9/11 attack attention had been drawn to Zacarias Moussaoui who had made suspicious comments during flying lessons.[37] He was interrogated, offered large cash awards, even a new identity and failing injected him with truth serums but he kept silent, leaving the FBI with the knowledge that an attack was imminent but no idea how or when. There was even a fear that a nuclear bomb had been stolen from a Russian arsenal raising the necessity of stopping a major disaster for millions of people. If one person knew how to avert the problem but refused to cooperate many would turn a blind eye to torture on the grounds of sheer necessity and may be even resort to torturing his wife and children. In the 1980s the Israeli government ordered a study of this issue, deciding that there was a hidden world in which the security services operated in the shadows and turn the proverbial blind eye. The state has a duty of safety towards its citizens but also maintaining human rights. Behind the scenes within the realm of national security systems, even with in liberal democratic states it is generally accepted that underhand policies are enacted, and immoral deeds such as the application of fear and torture may well happen. If the rule of law applies this should not be happening, as it would be like authorising a new Gestapo and buildings like Stalin's Lubyanka becoming acceptable. More to the point, if it is out of the hands of the political state turning a blind eye, it would soon escalate out of control by setting a precedent.

It is well known that waterboard torture has been applied even by democratic states such as America, and Guantánamo Bay has often been in the headlines under the suspicion of torture. There have been criticisms made of many governments of clandestinely moving prisoners to other

---

[37] See Kinsella, p.232.

countries where it is known that torture can be used to extract information, a form of illegal extradition (sometimes called rendition) for immoral reasons and mainly kept out of public view. This is occasionally raised and naturally denied, but the suspicions probably have some justification. The various secret services around the globe live under their circumstances in their covert world where they face immense dangers on behalf of their public, unseen and sometimes unrecorded. This dilemma should not be surprising in the world they inhabit when in their opinion they have to break normal conventions for public safety, but whether this is justified or not raises huge question marks and major dilemmas.

Most morally inclined people would object to torture but if it were a choice between one person and millions as in the ticking bomb situation many would waver in their convictions. It is a question which demands more thought as it raises the question of when necessity demands neglecting agreed moral standards. The ticking bomb syndrome is an extreme example, most people who are tortured are being forced to give information about others, organisations, and potential plans which although important cannot fall under the title of sheer necessity for saving a nation.

Military necessity

If a commander or even a foot-soldier is convinced that an illegal or immoral action would end the war or open the way to victory the same question of necessity and justification arises again, always recalling that the argument for necessity is casting any just war theory aside. For the ordinary combatant decision in the heat of frontline battle there is no time for philosophical discussion, but this is different for top commanders who must plan and consider all the ramifications. It can lead to disastrous outcomes at a tactical and moral level, and it applies not only to treatment of the enemy but to one's own troops.

General Mark Clark had convinced himself that the Italian War (1943-5) could be won if his troops crossed the Rapido River and although his subordinate officers carefully pointed out the inherent dangers, he issued the orders. In a brief action this cost 1,600 American soldiers their lives, 700 were captured, and only an estimated 65 Germans lost their lives. In the immediate postwar years this action went as far as a Congressional Hearing in which Clark was vindicated. Most observers since have seen this as a gross tactical error, he had been warned by officers on the spot, his belief that the crossing at that time was hardly an extreme military necessity, his critics claim he was driven by the need for personal success. The question rarely raised was how far any commander should consider the moral implications of deliberately sacrificing his men unless it is certain that it will create a momentous victory not just on the local battlefield but help end a war. Clark is a mere illustration of many such incidents throughout history. Commanders have an immense moral responsibility not just in the way they plan their strategy, proportionality in the nature of the attack, treatment of enemy combatants, but the way they sacrifice their solders' lives, especially if their motives are self-serving. No doubt moral responsibility is taught at various staff colleges, but the reality of history indicates the problems when in the face of war and not theoretical discussions. However, many argue there may be times when sheer military necessity is confronted with breaking war conventions and moral norms.

When faced with a total war where the stakes are survival of states against a maniacal tyrant there may occur times when even the most morally convinced men feel obliged to override accepted war conventions. Michael Walzer reminded his reader that Stanley Baldwin was hardly a warmonger, but he stated in the case of potential bombing that if a 'man has his back to the wall he will use the weapon whatever it is'.[38] When faced with the dangers of Nazism whose victory would have brought sheer evil a country is obliged to use extreme measures to defend itself and others, and as such the British bombed Germany as it was the only offensive available. This belief also led to Winston Churchill authorising the order to sink the French Fleet at Mers-el-Kébir (3 July 1940) on the grounds that it might fall into Nazi hands, although this was likely it was not certain, thereby the British resorted to killing their own allies.

The order that 'no prisoners be taken' breaks the rules of conduct in war, but it has been argued that military necessity sometimes must override even this major war convention. In a commando raid or a deep penetration

---

[38] Walzer. p.251.

in a jungle area when enemy soldiers are caught, in an operation which is essential for victory, they cannot be taken with the attacking force, by tying them up or setting them free it could potentially destroy the attack mission, and so it has been argued that it leaves few alternatives but killing them.

Demanding that soldiers must be ruthless and show no mercy has been a fault many times in military history. It is natural that commanders must warn their troops to fight because their lives and success depend on their profession of arms, but war convention and proportionality are all part of accepted rules of conduct. Nevertheless, over exalting troops to fight brutally presents the danger of fanaticism which gives some personalities a *carte blanche* to break the rules of war. This policy made the Nazi SS troops the most feared, and their atrocities were well-known and led to many postwar trials of SS commanders and units. This issue demanding brutality from troops was not just a German problem, but it touches upon nearly every combatant force. During Operation Husky, the invasion of Sicily, the American General Patton was well-known for demanding too much violence from his men, telling them, often in vulgar language, to shoot the enemy whenever there was an opportunity, to show no mercy and *to kill the bastards*. It would be like a rugby coach telling his team to hit the opposition hard even if it meant breaking their legs. It resulted in two cases when two of his soldiers massacred POWs on the grounds that Patton had told them to kill everyone. Patton tried to turn a blind eye, but they were convicted but were soon serving on the frontline again. Patton was also well-known for slapping his own soldiers in hospital beds for being cowards. This brand of sheer fanaticism was embarrassing for the Americans and led to a time of Patton's personal isolation from the front-line. However, such are the exigencies of war, he returned to active duty in mainland Europe where he proved a sound antagonistic style of leader, but he lost troops in an ill-advised unsuccessful mission to rescue his POW son-in-law. Many German generals thought he was the best Allied general which hardly served as a recommendation even though he was a dedicated soldier, because he had overstepped the boundary of war convention and morality.

The question of military necessity can be found in many issues which will be surveyed in this chapter, including the bombing of civilian areas, using

WMD including nuclear weapons, of imprisoning civilians, the scorched earth policy of destroying land and food resources, suicide missions, and much of which arises in the following sections which explore human rights.

Human rights for combatants in suicide missions

Combatants have human rights as serving soldiers as do POWs, they are often neglected in the crisis of war, sometimes by their own commanders or by their enemy. If a soldier is asked to go on a suicide mission the normal practice is to make it voluntary, or at least warn of the dangers. Throughout the modern era there have been examples of suicide attacks, in Japan known as kamikaze, and more recently with modern terrorism suicide bombings. When given no choice some soldiers protest. The so-called discipline of obeying orders may be applied, and in 1917 there was a mutiny by French soldiers who realised that more fighting in the suicidal attacks they had been ordered to do would lead to their inevitable deaths. As recently as October 2004 in the war on Iraq some American soldiers refused to drive fuel trucks in unarmoured vehicles close to Baghdad calling it a suicide mission. They faced hearings for their breakdown of discipline. In military terms disobeying orders is a crime, but as the Nuremberg Trial later established, illegal orders defined as immoral do not count, and Allied war manuals had to be changed.

Sending a non-voluntary participant on an identifiable suicide attack is illegal and immoral. The Japanese military were known for suicide attacks, and their volunteers may well have felt the social even religious pressure to stand forward for such a task, as may the Waffen SS with fanatical Nazi prompting. This has been mentioned in terms of Nazi Germany's fight at Stalingrad when they were ordered to fight to the death, mass-suicide, but General von Paulus had the moral sense to ignore the order. Soldiers always risk their lives, they know they can be injured or killed, but must rest in the hope they will win or at least survive. If this were not the case volunteers would be hard to find, as every person needs hope. To tell a man he must die in a suicide mission can only be justified if he is a true volunteer, not ordered or unduly influenced. However, influence is easily exploited. One commander may say to a group of men:

This attack needs you as a group to carry out a dangerous mission from which you will be lucky to escape with your lives. You have been chosen because you have worked as a team and support one another, you are all dedicated to our cause, you are all courageous men as proved by past deeds of glory, and the nation needs you at this critical time. I know you will volunteer, but if any individual wishes to opt out of the team, that is your right, and please see me after this meeting, or walk out now.

Another might put it differently:

This is a dangerous mission and necessary, but I must tell you it has all the hallmarks of a suicide mission, and if any of you wish to withdraw, please feel under no obligation, it will not cause us to judge you in anyway and would be quite understandable, especially if you have families who depend on you.

Even in a liberal democracy built on freedom unfair influence may be used. The terrorist's main weapon is the suicide bomber who often is a young person coerced by the belief he is following Allah's wish and will be rewarded in paradise and 70 virgins at his disposal. This comes from the Koran, but the number 70 (sometimes 72) arose from the Islamic Holy Tradition, the Hadith.

Suicide missions are not necessarily just small groups of men, but senior command has sometimes resulted in major incidents becoming virtual suicide missions. History is awash with such examples, one being the poorly organised Dieppe Raid (August 1942) when 6,086 troops were badly landed in an enemy occupied coast of France where the raid has been anticipated through lack of secrecy. They were put ashore and within hours 3,623 had been killed, made casualties or POW. Another was the first armed British merchant cruiser *HMS Rawalpindi* stumbling upon German battleships (August 1942, *Scharnhorst* and *Gneisenau*) when Captain Kennedy announced they were going to attack. He was awarded, understandably a medal for bravery, but it raises the question whether he had the right to lose 263 men in what amounted to a suicidal attack. There would have been no opportunity to allow the crew to volunteer or not. These incidents reflect many more and often written about in heroic terms, but they were suicidal without consultation.

A combatant is told and knows that being a soldier in war is life-threatening, but they should not be commanded to go and be killed in a planned suicide mission, any more than they should kill the enemy who is surrendering, or deliberately kill innocent non-combatants, rape or pillage, and ought to treat POWs as they would be expected to be treated in reverse circumstances. The conventions of war are or should be a mutual code which ought to be respected if humans wish to keep any sense of morality or self-dignity.

Human rights for non-combatants

The avoidance of attacking the innocent, the non-combatants, as noted many times is one of the prime themes of a just war, but as war has developed in the modern era the death and casualty rate has increased exponentially as has the suffering for non-combatants. Long gone are the days when armies met and fought at a distant spot, in a modern total war non-combatants cannot help but be involved. As early as the mid-19th century the American Francis Lieber wrote that the unarmed citizen should be spared but also added the phrase 'military necessity' which implies the rules may be overridden if it determines a successful outcome. It is impossible to tell if the citizens of the enemy agree with their leaders, even some of the soldiers may be reluctant fighters for a cause, but the essential point is that a non-combatant can do no harm to their enemy's soldiers. It is again a matter of proportionality, and although some citizens may be killed it should be extremely limited and out of sheer necessity, often covered by the trite term collateral damage. The days when armies would agree to meet at a particular area and start the battle at an agreed time has long gone as in the modern era the loss of a battle or war can lead to annihilation or the total loss of human rights when facing an enemy like the Nazi regime. It invariably means an increase in civilian deaths who are generally at greater risk than the armed combatant, especially when the civilian becomes the intended target, yet it remains a fundamental principle of the just war theory and accepted war conventions to distinguish between combatants and non-combatants, and force should only be used against those using military force. It is often argued, as mentioned earlier, that the non-combatant status has changed in modern times, with workers producing food, resources, and arms for the combatant. It is a fraught area and increases as war becomes more violent and the question of some non-combatants being innocent will be explored in the final chapter. In war there are many grey areas, but by killing those who produce food which is an essential requirement across the globe is highly questionable. The other issue is that modern weapons, in some cases, are themselves indiscriminate, and international law has attempted, by strong consensus to outlaw such WMD. The tragedy is in modern warfare that what is considered an essential target for the military will, almost inevitably, kill non-combatants. In many more localised conflicts, usually of an internal nature, paramilitary and armed groups participate, and it is argued they break war conventions because they do not have the discipline of authorised regular troops, though in the heat of battle even the best troops can ignore government instructions.

Death from the air

The non-combatant is often killed by bombing raids, and the question must be raised as to how to avoid what has been clinically termed collateral damage. Since the end of WWII with its infamous number of innocent deaths it has been hoped that modern developments have improved the aiming and direction finders for the necessary military targets. This failed to work in the Vietnam war with the defoliating agents as it ruined human life in vast areas, and the 'precise missile' attacks in the war on Iraq could hardly knock on the door to ask civilian workers and typists please to leave, and mistakes based on poor intelligence are all too common. Nor is it easy to define a non-combatant when fighting in an area like Vietnam where uniforms were minimal and often the Vietcong were hidden amongst supporting civilians. The maintaining of war conventions is never straightforward.

One of the fraught judgements was the decision made in 1940 to bomb Germany in WWII, it reflected Baldwin's statement that when man has his back to the wall, he will use the weapon whatever it is, and the British decided that bomber command was the only way of stopping a German invasion. It was deemed a military necessity in a state of emergency for survival. It was argued that the prospect of being overrun by Nazism had to be resisted at all costs, even ignoring the basic convention of not killing the innocent. Indiscriminate bombing was initially forbidden which was an impossible demand during that period of time, but after the Luftwaffe hit Coventry, a town known for manufacturing and deemed by Kesselring as a tactical target, bomber command was directed at cities. Lord Cherwell, Churchill's adviser, suggested that although it meant attacking workers' homes which meant killing civilians, he thought many Germans would be homeless creating havoc. The lack of modern aiming technology meant inevitably killing innocent non-combatants even if unintended, and it is known that it troubled some bomber crews. These actions overrode the established rules of war, and from Germany to Japan conventional bombing and two nuclear devices killed over a million innocent people. It was a policy intended to break the enemy's willpower, it had the elements of revenge, but it stirred a sense of resentment and determination amongst the public, whether the bombing was in Britain or Germany. Bombing civilian areas amounted to terror bombing supported by many people and the military, but it was also divisive in some military quarters and among many civilians.

By the end of 1942 it was clear that there would be no invasion of Britain, and a year later it was equally clear that the Germans were losing the war. The continuation of attacking a civilian population it was later argued was to shorten the war, with Bomber Arthur Harris believing the bombing could achieve this victory. If the year 1940 bombing was the point of necessity to stop the invasion, by 1943 this was redundant as an argument, and instead of carpet-bombing, tactical attacks against oil-refineries, weapon factories and steel works may have crippled the enemy more quickly than attacking homes. The American bombing, flying by day made a singular effort to hit industrial areas, the British at night were less able and aimed at large cities. It could be cogently argued that after the tide turned against German military success that strategic bombing had become a policy of terrorism sustained until the end of the global conflict in August 1945.

What made bombing even more questionable for many was when innocent people who were not enemy subjects were killed in French towns, (even by Free French bomber pilots) during the 1944 *Overlord Operation*. Field Marshal Montgomery ordered the bombing of Caen (July 1944) as his plans for its early occupation had failed. Allied planes dropped messages warning to evacuate the area, (a population of some 60,000 people) and this was followed by 500 bombers attacking. It is unknown for certain how many people died in the town's destruction but has been estimated as many as 2,000, and this was just one of many incidents. The historian Antony Beevor was even noted in a newspaper for his controversial statement that the bombing of Caen was 'close to a war crime', controversial as the paper stated, but probably the unpalatable truth.[39]

Tactical bombing was aimed at military areas and resources such as oil-refineries, industry, docks, and ports was attempted by both sides for practical reasons. However, as a deliberate policy carpet-bombing cities became the norm, which, after the horrific bombing of Dresden (February 1945) near the end of the war, gave Churchill doubts about the policy and, unfairly, muddied Bomber Harris' reputation as orders came from politicians and Chiefs of Staff. George Orwell suggested it brought home 'the horrors of war to those who supported in from a distance'.[40]

All sides were involved in bombing, but the logistical power of the Allies made them the most prolific in this theatre of war. The whole concept of aerial power had emerged during the Great War, and had been enhanced by an Italian theorist, General Giulio Douhet in his book *Il dominion dell 'aria* (Mastering the Airspace). In this work he projected that bombing the enemy's industrial resources and even the homes of the workers would be a decisive action for an early victory. His motive was to avoid the attrition of WWI and shorten a war, but that failed. It has been argued that those working in military industrial areas cannot be described as non-combatants, but this is a highly dubious viewpoint. The same is often claimed about scientists, not least those who devised the atomic bomb. However, while it is easy to blame scientists the initiative came from politicians fearful the enemy may build one first. The peculiarity of history was the first nuclear device was used against the imperial threat of Japan. The irony was that the American fire-bombing of Japanese cities caused more deaths than the two atomic weapons, and it was used because it was

---

[39] Observer, *Record Column*, 7 June 2009. Commenting on his book D-Day: *The Battle for Normandy*.

[40] See Walzer p.261.

believed that the Japanese would not surrender. It was argued the Japanese still had an estimated two million soldiers and might use or abuse POWs in the event of a main land invasion.

The Japanese had been fanatical fighters and if one imagines oneself having fought through the battle for Okinawa (April-June 1945) with 80,000 American casualties, anything which would bring the war to an immediate halt would be welcomed. Yet as with traditional bombing it meant the death of thousands of innocent people. Hiroshima (6 August 1945) felt the impact of a uranium bomb, nicknamed Little Boy, and was followed by another bomb, nicknamed Fat Man and using plutonium on Nagasaki (9 August) a few days later. It broke all the war conventions in terms of deliberately attacking mainly civilian cities, many have argued that it lacked proportionality in bringing defeat at the expense of innocence on a massive scale. Neither Hiroshima nor Nagasaki were major military or industrial areas, just the normal city life. The sheer necessity of stopping war was the *raison d'être*, but moral questions arise more sharply over the necessity of the second bomb, increasing with intensity over the claim that the second bomb may have been an experiment with the plutonium element. These bombs introduced humankind to a new era of WMD and the possibility of humankind's total annihilation, which is why the sense of a common global morality remains so important. The indiscriminate killing of innocent people was seen as essential in the need for victory in WWII, but it totally ignored one of the basic demands of a Just War. During WWII there was an overpowering necessity of not letting evil win, even if it meant breaking moral rules and war conventions. Another rule which was pushed aside, and often unnecessarily was the question of human rights.

Death at sea

At sea non-combatants were vulnerable on liners, hospital ships and merchant craft. Merchant ships were vulnerable for attack if suspected of carrying resources for the enemy, most were vulnerable to submarine attacks and on some occasions a more morally inclined U-boat commander, if in an isolated spot would give the crew warning to abandon ship before the torpedo struck. There are also plenty of recorded times when the U-boat surfaced to ensure the lifeboats had water and food and gave direction to the nearest land, which obviously could not work in a convoy system with naval support.

Hospital ships were also vulnerable often by mistake, but it was agreed they should be left unharmed. The British sank the Italian hospital ship *Po* which happened to have Mussolini's daughter on board, but even he admitted to his son-in-law that Italian hospital ships had been used as cover to carry fuel to North Africa. In war mistakes are made, subterfuge and deceit work hand in hand.

Liners are large and unmistakable, and it was agreed that their safety should be assured, but not even this was easily ascertainable when it was known many had been converted to carry thousands of combatants. Again, looking through a periscope in murky conditions, spotting a gun on a liner's deck, with on-the-spot decisions by the U-boat commander can result in understandable errors. In the Great War the *Lusitania* was sunk by a U-boat in May 1915 and this was replicated in September 1939 with the *Athenia* both causing public outrage as being unfair, with Joseph Goebbels trying to blame Churchill for doing it for the sake of anti-German propaganda.

Submarine warfare was once proclaimed as unfair as it amounted to ambush but as all sides used this method it was accepted as a part of naval warfare. Considerable propaganda was used during and after the war against the German U-boats, with often false claims of brutality, and postwar films had U-boats shooting survivors in the water. This type of propaganda by all sides has a dubious reputation in terms of moral behaviour. There was only one known German incident of this type of behaviour which related to a Kapitänleutnant Heinz Wilhelm Eck of U-Boat-852 when he sank the Greek freighter SS *Peleus* in the South Atlantic. Later he offered the excuse that he did not want the enemy to find the debris and evidence of his whereabouts. In the Japanese war there is overwhelming evidence that Japanese submarine commanders shot survivors in the water, but this was exceptional with the other involved protagonists. As mentioned above despite the war there was considerable moral efforts to save survivors. Within the first week of the war the famous Kapitänleutnant Gunther Prien of U-Boat-47 sank the British cargo ship

Bosnia. He surfaced and fired at the vessel to stop the radio operator sending out a distress call, and he saw as the crew scrambled to abandon ship the lifeboat toppling and casting the men into the water. He assisted by putting them into the other lifeboat and contacted a passing tanker to rescue them.

Later there was a classic example of a sense morality being exercised at sea which went badly wrong. In September 1942, U-Boat-156 under the command of Kapitänleutnant Werner Hartenstein sank the liner *Laconia* This 1921 built luxury liner had been converted to war-work and carried some two thousand six-hundred passengers including Italian prisoners, Polish guards, British military personnel and women and children, left floundering in the mid-Atlantic near Ascension Island. Hartenstein having realised the error surfaced and started to pick up survivors. He radioed Dönitz for help, and either Hartenstein or Dönitz asked for French Vichy help. U-Boat-506 under the command of Wurdmann and U-Boat-507 under Commander Schact came to assistance, as did the Italian submarine *Cappellini*. The Vichy French ordered three vessels out for the rescue, and their cruiser *Gloire* picked up many survivors.

However, on Ascension Island the Americans were setting up a secret airbase called *Wideawake* and they became aware of the situation. Hartenstein had sent an open message asking for help promising he would not attack any ship which came to assist. A single B-24 Liberator under the control of a Lieutenant Harden from Ascension Island passed over U-Boat-156 and saw the Red Cross on the U-Boat's deck. He radioed back to Ascension Island for instructions, and a Captain Richardson and a Colonel Ronin ordered the destruction of the U-Boat. This was a questionable decision, but in the heat of war and the determination to sink U-Boats to avoid further attacks, as well as the need for the secrecy of the American presence on Ascension Island it was a complicated decision to make on the spot, and easy to criticise in hindsight.

As the plane re-approached the U-Boat with its bomb-doors opening Hartenstein refused his survivors' requests to open fire. His U-Boat was not hit but damaged, and he had no choice but to evacuate the survivors and submerge to depart. He left them with directions, food and water and advised them that help should be on the way. Nearly a thousand people survived because of his humanitarian actions.

The war with the U-Boat changed and Dönitz gave the order that it was far too dangerous for a U-Boat to be caught on the surface, and the direct order was to depart as soon as possible regardless of the plight of possible survivors. As a matter of interest Hartenstein ignored the order when later in 1942, he sank the cargo vessel *Quebec City*, and surfaced to give the survivors sustenance and directions

The Nuremberg Trials condemned Dönitz to ten years in prison but much of this was probably due to the fact he had been appointed Hitler's successor, and as the *tu quoque* (you also did this) argument, albeit it banned, was evident, and he had the support of Admiral Nimitz, and many British senior naval officers but was condemned to a ten-year sentence. Compared to air and land warfare, naval conflict, albeit brutal, was balanced in terms of war convention, attempted to be moral where possible, and was more like the old-fashioned chivalry than total modern war.

Death by armies

Michael Walzer in his book emphasises that the second principle of war convention is that non-combatants cannot be attacked at any time. He then related the story of a soldier who before throwing a bomb into a cellar during bitter house to house fighting first shouted to warn any hiding civilian family.[41] It was like a surfacing submarine to help survivors, a moral attitude but one which was potentially dangerous. Instead of an emerging civilian he may have been greeted by an enemy hand-grenade. Where fighting occurs amongst the civilian population the risks of causing deaths to non-combatants is always prevalent. When the British and Commonwealth forces fought the Germans and Italians in the deserts of North Africa fewer non-combatants were harmed, prompting Field Marshal Albert Kesselring to describe it as a clean war. There were few civilian places in the vast North African desert areas, so it was not a surprising claim. However, it is intention

---

[41] See Walzer

and motive which make a situation moral, and when the war moved into Italy with its civilian population totally immoral acts were in abundance. Over the ensuing years of the campaign from 1943 to 1945 there were endless massacres, one at Biscari where Americans killed POWS, the rest were various German units against civilians or partisans some of the largest massacres in order were: Boves, Lake Maggiore, Fosse Ardeatine, Guardistallo, Sant'Anna di Stazzema, San Terenzo Monti, Padula di Fuecchio and Vivca, with many smaller ones and others unknown. The victims were partisans, but civilian hostages and reprisals tended to be the dominant victims, and there is no excuse or necessity for such actions. If Kesselring could describe the North African campaign as clean it made the battle for Italy dirty in the extreme. There were more civilians in Italy than soldiers, but two soldiers equipped with machine guns can control several hundred civilians and hold them at their mercy. On both sides, but most especially the Nazi and Fascist elements had no concern about war conventions unless it impinged on them when captured; such is human nature.

It was not just massacres which offered instances of the derogation of human rights, but women were raped across Italy, especially by French deployed Moroccan soldiers (Goumier, commonly referred to as Goums) who were nothing less than mercenaries who fought on terms with a license to loot and rape.[42] Their reputation was so bad that at one time the French authorities authorised mobile brothels. The abuse of women and sometimes children was not confined to Italy, as many soldiers when winning and on the rampage can be dangerous to vulnerable citizens. Perhaps the most frightening for civilians were the Russians fighting their way through Germany, with such a sense of revenge for Nazi behaviour in Russia no woman was safe, and rape was widespread. This was not the heat of battle which often causes misjudgements, poor decisions, reactions, and immoral behaviour, this was human behaviour at its worse. There may be war conventions, a universal sense of morality but when Shakespeare wrote about unleashing the dogs of war, he was close to the truth about human behaviour during uncontrolled conflict. Some individuals when armed, are revengeful, angry, with the freedom of action which war often allows and behave worse than any other form of life.

---

[42] See Walzer p.133.

At the Nuremberg Trial of 1945-46 it was understood that moral considerations were often cast aside in the heat of battle, but obeying orders was no excuse if the orders were illegal. Following the trials some Allied national Field Instructions had to be quickly revamped in the light of this proposal as did attitudes towards hostages and reprisals. After WWII non-combatants still suffered despite the Nuremberg conclusions, both from indiscriminate bombing and orders which took little account of civilian safety. In Korea and later Vietnam, if troops found themselves under attack it became a common policy to call in artillery support or air attack to clear the way which not only killed non-combatants but made human life in some areas untenable, and in both conflicts massacres of the innocent occurred. Officers could argue that their men's lives came first and foremost, and a commitment to preserve the lives of the innocent seemed to prove impossible. Even high-tech advances can never avoid this problem even though the demand to take due care of killing non-combatants is critical. Atrocities and civilian deaths continue to this day unabated despite war conventions and international agreements which are constantly ignored. In Rwanda both Huti and Tutsi fighters would hide in refugee camps making the innocent vulnerable, such attacks were not creating collateral damage or unintended, because at times hatred and revenge can be out of control in humankind making such incidents intended.

Reprisals and hostages

One of the most abused conventions is that of reprisal which can be applied at the highest level of command down to individual incidents. When a country is blockaded or bombed it is very normal for the other side to announce some form of reprisal which gives a legitimacy for illegal behaviour, but the question will be raised as to the alternatives. It has been proposed that the threat of reprisal can break the chain of action and deter wrongdoing. Michael Walzer in his work related that the Germans were executing FFI (French Forces of the Interior) who wore insignia to signify they were legitimate combatants. The French informed the Germans that if this continued, they would execute German POWs and on having no response executed 80 Germans. This action promptly gained cooperation from the German powers in that area. It appeared to have worked, yet the French had signed the Geneva Convention of 1929 which had re-affirmed barring reprisals against POWs. It could be argued that it worked, and as in domestic law guilty people are punished to protect others, this form of reprisal deterred further illegal acts, but it took an illegal act to work. Others return to the question of proportionality and ask if the French had killed five or ten would it not have worked just as effectively. Others argue about the gross seriousness of the case and that it demanded extreme reaction, but whether it is justified to meet abuse or evil with the same means is a tenuous argument on the grounds that it is simply wrong to kill innocent people. The Geneva Convention of 1949 followed up this issue with POWs, the sick, shipwrecked, wounded, and civilians in occupied territories, barring them from being reprisals.

Throughout WWII and beyond the question of reprisals and hostages remained a serious moral issue. It was deemed necessary when fighting partisans and resistance fighters when their family members were identified or civilians of their home community were held as hostages and then shot as reprisals, and all sides found reasons to use hostages as potential reprisals. This made civilians a significant resource for the enemy, (especially against what the perpetrators considered illegal such as resistance fighters) even though it was against war convention and totally immoral wherever it happened and for whatever reason, it also created a vicious circle. It remains a problem to this day, not least when terrorists take hostages with the threat of reprisal if their demands are not met.

Guerrilla war

They can be named resistance fighters, partisans but at one level or another they operate in what has been called guerrilla warfare in their resistance to a military occupation. Guerrilla fighters tend to be more political and ideological, resistant fighters and partisans are equated with freeing their country, but the distinction is often elastic and blurred. The terms used to describe this sort of fighting may be modern, but its history goes back to the earliest of times. For the sake of easy convenience, the term partisan shall be deployed to cover all areas.

It is a form of irregular warfare and usually occurs when a country has been defeated and in occupation. They normally wear no armbands or any form of insignia and this tends to eradicate any rights on the point of capture or surrender. They can hardly be regarded as traitors, and it could be argued that although the state has surrendered the citizens can still resist. On the other hand, many argue resistance is subversive both to the occupiers and their own government who surrendered, and when the partisan disappears into the civilian population there is the danger a whole community could be punished for harbouring a criminal.

On the other hand, there could be a claim for rights because partisans are defending their country where their army failed, and it has become a people's war. Often partisans initiate an incident to stimulate a severe reaction by the occupiers thereby infuriating others in the population to active resistance, which was often a ploy amongst communist partisans in France and Italy. Partisans are rarely able to release prisoners or hold them captive and stand in danger of breaking basic conventions for POWs. Where the partisan band grows, as in Tito's Yugoslavia they started to wear insignia giving them a sense of solidarity but greater vulnerability. In the IRA conflict in Ireland active members would appear at funerals or occasions when it was deemed safe, but soon disappear back into the civilian population. The key issue is that in such occasions civilian dress is their best disguise. When in WWII the Germans claimed hostages and reprisals for partisan attacks the question the public often raised was whether this action could justify partisans and their work. The Germans could claim it was a matter of military necessity, civilians could understandably argue that both sides were acting in an unjustifiable way putting innocent lives at risk.

It is not that an assassin can be always identified as wrong, anyone managing to kill Hitler and his henchmen in 1942 would have been deemed heroes. Wearing or not wearing an insignia may be critical for the occupiers, but by their very nature partisans must fight amongst civilians. It has been pointed out that often partisans are protected by the civilian population, but there have been times when the same population has lived in fear of the partisans who needed their food, resources, and cover. For the invader or occupier, the fear or annoyance is that somewhere in the civilian population they are hiding the fighters which blurs their vision of who is the enemy.

In Vietnam the American forces faced the same dilemma as it was known that villages hid the fighters and often the weapon caches with the probability that some of the villagers were the enemy. Sometimes helicopters were used to issue a warning of incoming shells or bombs to a village area to warn them, but such was the popular support for the Viet Cong this caused difficulties. As a result, the American policy was to uproot a village and place it elsewhere, often in unpleasant holding areas. It also led to the ruination of the rural way of life and culture, not least when defoliation bombs were used, and many fled to the cities. The detached citizen wanting to work and live blamed the partisans and guerrilla fighters, most the Americans.

It is an area fraught with moral difficulties because of the nature of the war being fought. There are those who argue that fighters who wear no insignia, who hide and use the civilian population are wrong from the traditional war convention, and because they put innocent lives at risk, they are immoral. Others that their cause was just and as civilians they had to right to fight back against oppression and had general civilian support. Whether the Americans were right in their involvement in Vietnam's disputes has divided American opinion since the war started. From the Just War theory, it is wrong to kill innocent citizens which this type of war virtually guarantees, on the other hand it is usually the civilian who is doing the fighting for the survival of his own home and territory and could be regarded as a form of self-defence. The more this writer considered the nature of this type of conflict, having read the views of many others, left him in total confusion, erring on the side of the partisan with the proviso that there had to be a chance of success, and it was genuine and justified form of self-defence.

Sieges and blockades

Sieges and blockades are a question of war involving life and death. All sides have tried blockades, in the Great War the blockade of German ports had severe effects, and in WWII the U-boat warfare meant the possibility of serious shortages for Britain as an island nation. Franco the Spanish dictator remained reluctant to join the Axis in WWII, not least because Spain was impoverished and weak after the Civil War, but the fear of a British blockade against essential imports would have led to disaster. Blockades are enforced sanctions but whereas sanctions might lead to war, blockades are a form of war.

The siege is the oldest form of warfare and always involves civilians. In any siege soldiers are first to be fed and the trapped civilians are in danger. Michael Walzer takes us back to the siege of Jerusalem (AD 70) where the intention was to attack the inhabitants because victory was necessary. In the 12th century Jewish law (Talmudic) stated that a city should be surrounded only on three sides to allow those who wanted to escape to save their lives.[43] Assuming, one must conjecture, that this would be watched to ensure it was a one-way traffic system. The Jewish Talmudic law represented the right of the innocent to leave. In any siege non-combatants will be present, and it could be argued they are part of the defence if they dig trenches and reinforce walls, whether voluntarily to defend their homes or obliged by the combatant defenders. The population of non-combatants may have increased as they may initially have regarded the city as a refuge from invading forces.

More civilians died in the siege of Leningrad (September 1941- January 1944) than died in the bombing raids on Hamburg, Dresden, Tokyo, Hiroshima, and Nagasaki taken together.[44] It was the most damaging siege in history, people were arrested for cannibalism, and although the figures remain uncertain it has been estimated at one stage 100,000 people died every few months from German shelling and sheer starvation when at its worse, and the final death toll was between 600,000 to a million people, some believing the true figure was closer to three million. This writer visited the city in 1970 and saw the mass graves where residents still placed sweets on the huge grass mounds where they speculated their loved ones might be, all against the background sound of mournful music.

The question of the right to leave such a disaster is often raised. It has been argued that non-combatants should have the right to leave. If the commander of the city wanted to drive them out to make more food available for the military garrison it was considered legal to drive them back in; in Leningrad the Germans certainly blocked such retreats. At a later trial the judges did not appreciate this argument but accepted its base lines. Michael Walzer pointed out that the dangers and complexities were immense, ranging from coercion of non-combatants by the home

[43] Walzer p.168.
[44] Walzer p.160.

garrison, whether they consented to be defended, or coerced by their attackers, their homes became targets for shelling, and if they managed to leave it was fraught with danger, a totally unsafe situation. It has been noted that while civilians died from starvation in their millions, soldiers survived on their food rations. In any siege when it is decided to cut off water, power supplies and block medical aid it has the well-known double effect argument that it is done to achieve victory knowing it will kill the innocent, including women and children.

Long after WWII serious sieges emerged in the civil war of the mid-1990s with the Bosnian-Serb attacks, and with lightly equipped UN forces merely being able to observe, the destructive attacks continued. There were no safe havens for non-combatants as it was an ethnic cleansing war, and the object of the attack were citizens and their combatants. Shells and mortars deliberately targeted homes, hospitals, and schools, and as is known in Sarajevo snipers made a point of targeting any non-combatant caught in the open as matter of mere routine. The genocidal nature of such wars totally ignores the rights of the innocent and for many demands a UN intervention and not just observers.

A siege coupled with a blockade and scorched earth policy can place a whole country into a state of serious siege. The British naval blockade of Germany caused serious problems for Germany in the Great War. Michael Walzer described it as the doctrine of double effect because its intention was to reduce Germany's military capability but with the knowledge that non-combatants would suffer. It starts as an aggressive form of economic warfare and inevitably develops into full-scale warfare.

Sanctions

Sieges and blockades invariably involve innocent non-combatants, whereas it is often argued that sanctions against another country can produce the same effect but do not usually amount to a life and death situation, but some argue sanctions can create the same effects as a siege. When the League of Nations announced sanctions against Italy for invading Abyssinia (1935-6) for Mussolini it was more matter of political annoyance for being treated as a pariah state, suffering economically with a shortage of coal but for a brief time until Germany came to the rescue. Sanctions are generally not seen as a cause of war but there is always the danger that they may prompt conflict. A sanction is often regarded as a political or diplomatic way of either punishing a wrong or trying to persuade the target to a change of mind or attitude. The political world creates sanctions and it is the same problem as with war, bearing in mind the intentions and motives behind the decision making.

Technically a sanction does not challenge the sovereignty of a state, so when America withdrew trading arrangements with China because of their violation of human rights it was an economic challenge, an American right trying to persuade China to change its ways. It was different with the Cuban crisis (1962) as it sanctions amounted to a blockade which was close to declaring war. This form of diplomacy is often suspect, and even when supported by international agreement as with Italy in 1935-6 is not always effective, but it remains an alternative to warfare. It has been proposed that five criteria should be applied before the use of sanctions.[45] First they should be a response to an obvious act of injustice, secondly, there should be a reasonable chance of success, thirdly, only after other means of persuasion have been tried, fourthly, sanctions should not abuse fundamental human needs such as medicine and food, and finally the argument of proportionality. 'The great difficulty here is that economic sanctions are a blunt instrument for achieving desired political reform; their effects almost invariably fall indiscriminately upon the population of the targeted state'.[46] Walzer also raised similar criteria over imposing sanctions, nevertheless, this debate raises the questions if the innocent should suffer because sanctions generally ensure they do. On the other hand, many believed that when Iraq occupied Kuwait insufficient time was given to sanctions which may have avoided the deaths caused by Desert Storm, deploying sanctions as a non-lethal indicator of the political will. For many sanctions are an attempt to avoid armed conflict but to indicate a horror at another country's behaviour. At the time of writing Russia has attacked Ukraine (2022) and various financial and economic sanctions have been applied against the Russian Federation to indicate the feelings of other

---

[45] See Kinsella, p.279.
[46] Kinsella, p.279.

countries, in this case the sanctions can apply both ways, and the world is currently in a state of economic shock, and wondering whether the mentioned threat of nuclear war is no longer anachronistic as the major powers watch this delicate situation, with everyone hoping that sanctions, military miscalculations and political mishandling do not lead to a global disaster.

As is the usual dilemma with humankind no one can predict the outcome of human behaviour. If a man is barred for a time from a public house for causing a fight he may return a reformed person, or he may set light to the pub one night. Holding back a troublemaker may be the right thing to do, but it might create trouble for others who are innocent and therefore needs sound and moral reasoning as in any proposed restrictive action. It is generally accepted that crimes such as abuse of human rights, mass rape, genocide, ethnic cleansing, apartheid are wrong, but there are areas which for one society's views are ethically sound, for another somewhat dubious. The clash with religious faiths and ideologies are areas which tend to be grey and somewhat elastic. It is easier to evaluate a problem when natural justice finds an issue clearly repugnant. Boycotting goods from a country which uses slave labour is one thing, but in doing so because another person's faith allows several wives and permits girls under 16 to marry is another.

In war a fundamental theme of war convention is that innocent people are not harmed, this must also apply to sanctions. If sanctions impinge upon foods and medicines and create impoverishment it will hit the poorer people first and most harshly. In extreme causes it would affect the whole population, and not only create severe suffering but lack of medicine and nutrition could mean death. Those who are well-off, usually including the national leaders will suffer less, making as Mahatma Gandhi once stated that 'poverty is a form of violence', which is an often-ignored view but close to the truth. Precise considerations must be scrutinised before a sanction is set in place. Questions which need to be asked are whether it is a single tyrant or evil government, and whether the wider population agree with their leadership as can happen with such crimes as ethnic cleansing and religious bigotry. Other questions are whether sanctions will avert war and have humanitarian requirements been considered.

When checking this area of sanctions, it transpired that the UN Security Council in several cases used sanctions where there had been wrong-doing (Somalia, the old Yugoslavia, Haiti and so forth) and other countries have also done this with America making the most sanctions of any other state. It has been cogently argued by the writer Joy Gordon that sanctions 'cannot legitimately be seen merely as a peacekeeping device, or a tool for enforcing international law...rather, I will suggest they require the same level of justification as other acts of warfare'.[47] This same writer described sanctions as a bureaucratised form of siege warfare which would be correct when used for the wrong political reasons, and because sanctions are a form of siege warfare they mean that death is slower than in a traditional war, and must always come under the scrutiny of just war norms which is not unreasonable.

---

[47] Kinsella, p.303.

Chapter Eight - Post War Justice
Jus post Bellum

Introduction

When the Normans successfully invaded England in 1066, they took ownership, there were times of friction, inequality, and over the centuries who was Norman, Saxon, Viking, or Jute slowly became irrelevant as all became English, and any DNA test today would be difficult if not impossible to isolate any one race or tribe. Other conquered nations became part of another country, sometimes forgotten, other times still seeking independence. Sometimes punishment was given out to the conquered leaders, the citizens treated as slaves, and there were no realistic rules or laws to apply, just the whim of the conquerors who administered their own victor's justice. Following the Napoleonic wars there were no trials, but Napoleon eventually landed up in St Helena and this was followed by nearly a century of general European peace and stability, with a few interruptions like the Crimean war. It took to the 20th century until trials of a legal nature were considered, first suggested following the Great War 1914-18 but with major significance following WWII because of the political and military iniquities which had shaken the world.

Nuremberg trial

There had been a precedent of putting war criminals on trial, in the modern era there was the case of Breaker Morant, who was an Anglo-Australian drover, poet and military officer convicted and executed for murdering six prisoners of war and three non-combatants in two separate incidents during the Boer war.

However, at the international level, after a hundred years following the Napoleonic wars with its fairly peaceful resolution, there was the Great War (1914-18). It concluded with a demand for national retribution, when the recriminations of the Treaty of Versailles gave the 'Germans a near universal agreement that such treatment was unjust and intolerable, making the Versailles Treaty perhaps the only political issue, around which there was widespread agreement in Weimar Germany'.[48] This same treaty set a precedent for a major trial relating to causing war and war crimes, and a list of nearly a thousand criminals ranging from the Kaiser through to the top military commanders to officers was proposed. The Germans were angry at this proposal, and the Allies agreed the Germans could hold the trials in Leipzig resulting in thirteen convictions causing a sense of embarrassment amongst the Allies.

This undoubtedly led to divisions of opinions as to a trial after WWII. Churchill at one time proposed to take the main outlaws as he called them, about 50 to a 100 people, and shoot them after rapid Court martials which would have amounted to kangaroo courts. Lord Simon, the Lord Chancellor agreed with this policy, Anthony Eden claimed that the Nazi war-machine was beyond the scope of normal judicial process, and even the then Archbishop of York agreed.[49] The British believed that minor criminals should be dealt with by those countries where the offences had taken place. It was as early as autumn 1942 that the Russian Foreign Minister spoke to Eden about a postwar trial, with most remembering Stalin's brutal purges of the 1930s which raised suspicions, and not helped when at the Tehran Conference Stalin had proposed executing some 50,000 to 100,000 German staff officers to which Churchill strongly objected. Even near the end of the war Churchill still believed in a summary execution of the main leaders, but this may have been his concerns over the legalities of different nations in such a trial resulting in confusion and even creating unnecessary hostility. It was the American Henry Stimson, the Secretary of War, who demanded that the due process of American law should be upheld and had Roosevelt's support. Churchill proposed the defendants should include Japan and Italy, but the former was considered a separate issue by the Americans and Russians, and in the initial drawing up of proposed names there were only eight Italians named, so it was put aside with the possibility of further investigation.

---

[48] Fulbrook Mary (Ed), *Twentieth Century Germany* (London: Hodder Headline Group, 2001) p.23.

[49] Overy, Richard, *Interrogations* (London: Penguin Books, 2002) p.8.

It has long been understood that any legal procedures should not have any political interference, but under the circumstances it proved an impossible demand, and later despite protests from its Allies the USA made a political decision not to allow the Japanese Emperor to stand trial on the grounds of postwar reconstruction, despite the argument that justice cannot be served if politics intrude.[50] In terms of a valid jurisdiction, the Allies were the only legitimate authorities in Germany and the International Military Tribunal was deemed legitimate as there was no other form of government.

The intentions and nature of the Nuremberg Trial varied not only between the Allies but even within every legal team. The legal and moral attitudes differed and included elements of revenge borne of the hatred of the Nazi behaviour, some as a form of education both for the German people and the world, a cleansing exercise, others saw it as having national motivations, but with the insistence of the Americans the trial was proposed to be based on acceptable lines of appropriate judicial proceedings and it was backed by enough evidence to fill a library. After much debate, the Allies agreed on four distinct indictments. The first was participation in a common plan or conspiracy for the accomplishment of a crime against peace. Secondly, planning, initiating and waging wars of aggression and other crimes against peace. Thirdly participating in war crimes, and finally, crimes against humanity. The trial was deemed essential because Nazi behaviour, led by Hitler, had produced a horrific total war with not only every conceivable form of crime but with new ones, some worse than the most barbaric times in recorded history, and much of this had been planned or considered before 1939.

The trial had its critics at an international level, not least that the Soviets sat in judgement with many knowing about the Russian massacres at Katyń, and joining with the Nazi regime to occupy Poland. Another was the banning of the *tu quoque* argument (you also did the same) which had massive ramifications, not only in terms of the Soviet presence, but in matters of conducting the war at sea, on the land, and from the air, not least on the fraught question of obeying orders, hostages, and reprisals, and many of the Allied Field Manuals had to be rapidly changed, making the

---

[50] Kinsella, *The Morality*, p.345.

victorious Western Allies reconsider some of their own rules of engagement. The American and other Field manuals were changed about taking hostages, and it was stipulated that American and British soldiers should only obey lawful orders.[51] The British Manual at paragraph 454 had once stated that 'Reprisals are an extreme measure because in most cases they inflict suffering upon innocent persons. In this, however, their coercive force exists, and they are indispensable as a last resort', the argument of military necessity.[52] Nuremberg's revelation of atrocities against civilians caused the democracies to rethink their military policies. For many citizens across the world these technical issues were of no concern, as they wanted a ruthless enemy who had carried out totally immoral and evil deeds convicted.

There is insufficient space to cover the trial which can be read in a brief book by this writer, *Blind Obedience and Denial*.[53] However, it should be noted that obeying orders was a critical issue which caused deep concern. Every nation has traditional values which often differ from their neighbours, the United States and China and many others have capital punishment whereas in many European countries and elsewhere it is illegal, however, nearly all know countries expect orders to be obeyed especially by their military. The defence of obeying orders was used in the mass genocide called the Holocaust and other mass murders, and the Nuremberg trial arrived at the policy of illegal orders which for many military men was *ex post facto*, a law that retroactively changes the legal consequences. It is a deep human instinct to obey orders often without question. A psychologist called Stanley Milgram carried out experiments with electric shocks demonstrating that people would inflict powerful electric charges on others, because an authoritative figure informed them it was correct to follow orders.[54] It was and remains a critical human issue which crosses all national borders.

---

[51] See Neave, Airey, *Nuremberg*, ((London: Biteback Publishing,1978) p.372.

[52] The UN War Crimes Commission, (1949) *Law-Reports of Trials of War Criminals*, London; Volume VIII.

   Case No 47, London: *Law Reports of Trials of War Criminals* p.12.

[53] Sangster, Andrew, *Blind Obedience and Denial* (Oxford: Casemate, 2022)

[54] Milgram Stanley, *Obedience to Authority* (New York: Harper and Row, 1974).

Critics focused on the claim the IMT (International Military Tribunal) had no jurisdiction and tended to apply *ex post facto* law, adding that all sides were guilty in one way or another. It has also been pointed out that the legal demand of *nullum crimen sine lege* (no crime without law) had not been observed. It has been further argued that the Nazi regime was immoral, but not all moral laws are necessarily illegal, unless like Iran where there is a Morality Police Force. The issue of *mens rea* (intention) and *actus reus* (the guilty act) had to be established, but this must come from the top echelons and again, in the lower ranks should include knowledge of illegal orders which was new to most participants. This meant that the Nuremberg judges had to set two major principles within the same bracket, namely that individuals ought to be responsible for their own behaviour not using obeying superior orders as a defence, and yet being part of the Nazi regime was not in itself a criminal act.

Various churchmen across the globe were critical, many of whom pointed out the hypocrisy following the use of atom-bombs, some calling it a show-trial, others that it was victor's justice. A Senator Robert Taft regarded the trials more as 'an instrument of government policy, determined months before at Yalta and Teheran'.[55] The claim of hypocrisy has often been levelled at the Nuremberg trial often based on moral grounds, Niall Ferguson writing that 'they were accused, firstly, of the planning, preparation, initiation, or waging of a war of aggression, or war in violation of international treaties', which was a common problem or habit of many nations.[56] Hypocrisy was a well-known theme and often justified, not least when the Soviets presented highly dubious evidence claiming the massacres at Katyń were carried out by the Germans, when many in the courtroom knew it had nothing to do with the Nazi regime.

Had the *tu quoque* argument not been banned there may have been less hypocrisy and Admiral Dönitz may well have been found not guilty. Telford Taylor, one of the prosecutors later wrote that in such trials the rules should apply to all sides: 'I am still of that opinion. The laws of war do not apply only to the suspected criminals of vanquished nations. There

---

[55] Davies, Norman. *Europe, A History* (London: Pimlico 1977) pp.1054-5.

[56] Fergusson Niall, *The War of the World* (London: Allen Lane, 2006) p.579.

is no moral or legal basis for immunising victorious nations from scrutiny. The laws of war are not a one-way street'.[57] It was impossible to keep politics out of the legal system, As Richard Raiber, who researched the crimes of Field Marshal Kesselring wrote, 'Unpalatable though it might be to many…the twin doctrines of military necessity and national sovereignty will continue to erase all potential laws of war'.[58]

Hypocrisy is a powerful aspect of humankind's nature, but it was not the reason for banning the *tu quoque* argument which claimed that an illegal action cannot be avoided on the grounds that another also did the same thing, but it certainly presented some embarrassing moments. The popular historian Max Hastings noted that when General Patton did his best to ignore his soldiers massacring Italian POWs in Sicily, wrote that 'Patton, whose military ethic mirrored that of many Nazi commanders' claimed these killings had been thoroughly justified.[59] In terms of war conduct all the protagonists had committed various levels of abuse of accepted war convention.

Obeying orders was a common defence utilised at the highest level by senior military such as Keitel and Jodl. This was followed by using Hitler's influence, others tried innocence and ignorance of what happened, some blamed others with Hitler, Himmler, and Bormann at the top of the list and conveniently dead, forgetfulness, amnesia as with Ribbentrop, some lied or presented misrepresentations, Göring tried to justify his actions, and military men like Dönitz and Jodl argued duty.

In terms of the value of such a major trial various opinions have surfaced over the years. The British prosecutor Maxwell-Fyfe reflected indirectly on Augustine's demand for peace writing that people wanted peace and that 'most men at the close of the war wanted a better world.'[60] The main trial ended in 1946 but trials continued in the West with the Subsequent

[57] Taylor, Telford. *The Anatomy of the Nuremberg Trials* (New York: Alfred Knopf, 1992) p.641.
[58] Raiber Richard, *Anatomy of Perjury* (Newark: University of Delaware Press, 2008) p.184.
[59] Hastings Max, *All Hell Let Loose* (London: Harper, 2011) p.445.
[60] Cameron J, (Ed) *The Peleus Trial* (London: Hodge, 1948) p.xiii.

Nuremberg Proceedings (SNP) until 1949 conducted by the Americans with the British holding further trials, not least in Italy, the most notable case being against Field Marshal Kesselring.

The trials had anticipated that the exposure of evil of the Nazi regime would be instructive in helping the German public realise the true nature of the Nazi regime. How much the German public knew remained a contentious issue, but in late 1945 they were shown a newsreel film *Die Todesmühle* (The Mill of Death), where most sat in silence 'without visible emotion. Some women wept; others laughed hysterically, then burst into tears, men were seen sitting with bowed heads, covering their faces with their hands'.[61] Whether the trials were able to probe that the problem was more political than military remains contentious, this writer believes the former carried the major responsibility in the overall scheme of events.

Post Nuremberg

Whether the trial was meant to educate the German public remains a complex matter. In the late 1940s the German public had been torn apart and bewildered as the rest of Europe, divided between victorious Allies, starving, humiliated, the Nuremberg trial was for the public an event taking place beyond their interest as survival was the key to their future. To stay safe, it was critical for individuals and families to distant themselves from any Nazi past and simply survive. For a time there followed a denazification process where the innocent had to convince the local committee that either they had never been Nazis, or if they had been Nazi Party members it was because they had little choice, while others argued that they did nothing wrong but made the wrong choice in thinking Hitler had promised peace as well as restoration of Germany. Even those released in the major Nuremberg trial had to go through this process.

Because of the nature of the Cold War and the need to keep West Germany onside, the SS and fervent members of the Party's top leaders were allotted the major blame. The collective memory is easily changed and

---

[61] Manvell R & Fraenkel H, *The Incomparable Crime* (London: Heinemann, 1967) p.244.

manipulated, and the Cold War was part of this process. This issue raised the phenomenon of *Vergangenheitspolitik*—the politics of memory, a word first used by the German historian Norbert Frei. As the Cold War became more serious the British and Americans deliberately started to refer to the 'image of a clean Wehrmacht', as it was clear that West Germany could act as a buffer state between the Soviet world and the West. With the later trials of commanders such as Kesselring and von Manstein it occurred to the Allies that it would not be helpful if admired German generals were under death sentences or interned. Very soon after the major trial the Western powers 'colluded in the fiction of a decent but simply soldiery led astray by Nazi ideologies.'[62] This reflected sheer ignorance and wishful thinking, particularly in the cases of officers like Manstein and Kesselring, but compelled by the sudden need to bring West Germany into the new NATO alliance. Thus, despite the Tribunal not finding the OKW a criminal organisation, it presented Eisenhower with the political opportunity to claim that 'the German soldier as such had not lost his honour' the despicable acts had been committed by a few; indicating the impact of the Cold War was mounting.[63] Despite the immediate Allied injunction against Germany possessing arms, 'the mounting fear and hostility towards the Soviets convinced Western Powers of the Federal Republic's strategic importance as a military force on the European continent'.[64] Once again, highly contingent imperatives of political expediency clashed and overcame the supposed 'independence' and 'universality' of liberal standards of law and justice.'[65]

This political situation gave a significant trait to the collective memory but came under renewed scrutiny when an exhibition was presented under the title of 'War of Annihilation: Crimes of the Wehrmacht 1941-44', by the Hamburg Institute for Social Research 1995-1999 and was deeply

[62] Burleigh Michael, *Moral Combat* (London: Harper Press, 2010) p.550.
[63] Bartov, O., Grossmann, A. and Nolan, A. *Crimes of War* (New York: The New Press, 2002) p.171.
[64] Hébert, Valerie Geneviève. *Hitler's Generals on Trial* (Kansas: University Press of Kansas, 2010) p.4.
[65] Salter, Michael. *Nazi War Crimes, US Intelligence and Selective Prosecution at Nuremberg* (Abingdon: Routledge Cavendish, 2007) p.7-8.

controversial as it proposed shattering the myth that the SS were the only culprits.

On a wider view the Nuremberg trial established a blueprint for the future. It was used as a basis for the Eichmann trial, and for international courts at The Hague when dealing with the Balkan wars, and at Arusha (Tanzania) for the genocide in Rwanda. It helped develop international law and was the precursor for the 1948 Genocide Convention, the Universal Declaration of Human Rights 1948, the Nuremberg Principles 1950, the Convention on the Abolition of the Statute of Limitation on War crime against Humanity 1968, and the Geneva Convention on the Laws and Customs of War 1949 with its supplementary protocols of 1977. The Nuremberg Trial helped establish a permanent International Court by being the first one of any substance.

The trial had made a major effort to explore the war years, with documentary evidence, witnesses, films, confessions, and diaries leaving no doubt as to what happened, which vilifies those individuals known as Holocaust deniers. In many ways the trial was politically necessary to show the dangers of authoritarian rule leading to the lowering of moral standards as epitomised under the Nazi regime. However, new identities emerged claiming the Germans were victims of a regime, of disastrous defeat, national division, and of arbitrary Allied policies. Nuremberg and the legal precedents it established have defined behaviour before and during war, but sadly has not changed human conduct.

Japan, the Tokyo tribunal

As mentioned in Chapter Two the postwar trials in Japan were different from Nuremberg in a variety of ways. They held in common the preceding debates as to who should be put on trial between the various Allied countries called together by the Americans when General MacArthur created the International Military Tribunal for the Far East (IMTFE) with a charter close to the Nuremberg pattern. As with Nuremberg (the site of Nazi rallies) so in Japan the significant building of the Japanese Army Academy in Ichigaya, Tokyo was selected. The Nuremberg indictments were also used, crimes against peace, war crimes and crimes against humanity. However, the crime against peace was the central prerequisite in the major trial, and other charges were heard outside Tokyo. The prosecution set out to confirm that the leaders knew atrocities had been committed and this trial took 192 days concluding on 27 January 1947, with seven defendants being sentenced to death, six to life imprisonment, and others paroled in the mid-1950s. There was a high degree of disunity amongst the international judges, and as mentioned in Chapter Two the Indian Justice Radhabinod Pal opposed the conclusions as did some French and Dutch judges, and it was criticised for being a white man's trial. There were subsequent trials held all over the Far East and IMTFE, often referred to as the Tokyo Tribunal sentenced and executed many war criminals.

The Japanese had signed the 1929 Geneva Convention on Prisoners of War, as well as the Convention on Sick and Wounded but failed to ratify the POW Convention. The treatment of POWs, the sick and wounded, as well as massacres of people, enforced prostitution were all mentioned at the trials and have been widely broadcast since through films, books, television documentaries, and many Japanese soldiers not only broke international but some of their own military laws. Some of the Japanese behaviour shocked the outside world, and as there were those who were angry with the Germans for allowing themselves to be taken in by Hitler and the Nazi regime, there some who thought the Japanese barbarities virtually made the Japanese a different zoological species. Whether beheading prisoners, or bayonetting them for practice, or burying them alive is better or worse than leading children into gas-chambers is not worth discussing, the whole WWII disaster in matters of evil behaviour is rather an indicator of how low humankind can sink.

The after-effects of the war and trials have been outlined in Chapter Two dealing with corporate memory and identity and continued into the current century. The Americans refused to put the emperor on trial, possibly based on the knowledge that many Japanese regarded him as almost deity. Emperor Hirohito (Shōwa) was the 124th emperor of Japan (ruling from December 1926 until his death in 1989), and General MacArthur recognised that if there were to be a successful postwar period putting the emperor on trial would not help establish stability. There was always the anticipated hope that Japan could be turned into a friendly state with the fear of communism growing by the month. This was seen by many as a form of political intrusion into the legal system, but from the pragmatic point of view it established a sense of peace which was the essential demand of St Augustine.

Trials in Italy

(Author's note: *this brief section on post WWII Italy was written at my request by the well-known Italian historian Pier-Paolo Battistelli. Italian politics past and present has always been open to sudden changes, and it seemed best to have the insight of an historian well-versed in Italy's recent history*)

Italy is perhaps the most representative case of how political needs prevailed over justice and the judiciary systems even in a liberal type of society. In early 1945 the situation in Italy was chaotic, with part of the country under Allied and some under German control and with a large fragmented partisan movement, with a past which was best forgotten, namely the war fought by Italy until her surrender in 1943. To summarise the situation the following were of postwar legal interest:

- war crimes perpetrated by Italians during the colonial war until 1940 (Libya, Ethiopia).
- war crimes perpetrated by the Italians during the Second World War until September 1943 (occupation of Yugoslavia, Greece, Soviet Union, and Libya).
- war perpetrated by the Italians against British and American soldiers until September 1943.
- war crimes perpetrated by Italians fighting alongside the Germans since 1943.
- war crimes perpetrated in Italy by the Germans since 1943.
- war crimes perpetrated by Partisans, Italians, and Yugoslavs, since 1943.
- war crimes perpetrated by Allied forces in Italy since 1943.

Such an impressive list would require several volumes for an in-depth analysis and would be revealing. Both Field Marshals Pietro Badoglio and Rodolfo Graziani fought in Libya against the Senussi insurgency and in the

Italian war against Ethiopia, in both cases committing evident war crimes, with Badoglio even boasting of it to the German Field Marshal Wilhelm Keitel. Badoglio, after replacing Mussolini in July 1943 took over the Italian government, he openly collaborated with the Allies and thereby escaped any extradition request from Ethiopia. Following his death, his birthplace even took his name: Grazzano Badoglio. The Ethiopians wanted to process Graziani as well, who was put on trial for having taken command of the armed forces under Mussolini's 1943-45 *'Repubblica Sociale'*. He was sentenced in 1950 to 19 years imprisonment, but he was released after four months and never faced an Ethiopian tribunal.

The Yugoslavian government wanted to put on trial General Mario Roatta, former Chief of Army Staff, and commander of the Italian Army in Yugoslavia, but Roatta followed the king in his escape from Rome. He was arrested and put under trial for his involvement in the killing of the anti-Fascist Rosselli brother and found guilty, but in the meantime he had managed to escape to Spain along with his wife. By the time of his comeback the Court of Appeal had overturned the verdict, which allowed Roatta to be reinstated in rank and pay.

The extent of war crimes committed by the Italians was extensive and can be compared to those committed by the German troops given comparative numbers. However, most of the war criminals escaped justice (only a few were taken prisoner and sentenced in Greece and Yugoslavia) thanks to an elaborated political scheme which had the Italians being portrayed as victims of the German occupation. Thanks to absurdities such as the film *Captain's Corelli Mandolin*, the self-acquitting myth grew in time, crossing the borders of Italy. Cover-ups were the norm, and all with the complicity of American and British officials in the light of the emerging Cold War.

This complexity leads to the crimes committed against British and Americans. The notorious case of General Nicola Bellomo was curious. Bellomo was the local military commander in the Bari area in southern Italy, and he was involved in the killing of one British officer and the wounding of another. Bellomo, who gained fame for opposing the Germans and handing over Bari to the Eighth Army in 1943, was put on trial in July 1945 and subsequently executed. The Italians never considered

it a war crime and in 1951 posthumously awarded Bellomo the Silver
Medal. As shown in a historical essay, not only part of the records went
missing but lack of interest largely contributed to the trials conducted
against Italian war criminals being rapidly forgotten.[66] The same applied to
war crimes committed during the 1943-45 period, which involved both
sides since partisans as well (including Yugoslavs) committed war crimes
while the war was raging in Italy.

After the war the Milan Assize Court found that a large number of Italians
actively cooperated with the Germans, mainly for money and personal
gain. When it came to examining the role of those parts of the armed forces
or formations, excuses were easily found to lessen their involvement.
Consequently, those Fascists who had joined the 'Black Brigades', mostly
employed against the partisans under SS command, were to be considered
potentially guilty only if they had voluntarily and consciously joined these
formations. The clear intention of the courts was best revealed by the case
of Colonel Giuseppe Baylon, formerly Chief of Air Staff and as such
actively cooperating with the Germans. His alleged involvement in the
fight against the partisans failed to be proved, and his cooperation with the
Germans was explained as protecting his men. Baylon was found not
guilty, like many other high-level collaborationists.

In 2003 the Italian parliament set up the customary enquiry committee
(which in Italian terms is an option not to go too deep into the matter)
following the discovery in 1994 of 695 files concerning war crimes
committed in Italy by the Germans in a hidden archive. It was soon clear
that the Italian government decided not to proceed to avoid alienating
Germany at a time when she was needed to support the country financially
and economically. The farce of some trials being held in the early 2000s
against people aged 80 or more, none of whom were imprisoned, revealed
the sorry state of the Italian justice system.

The most infamous war crimes perpetrated by the Allies in Italy took place
in Sicily by American hands, but it seems unlikely that anybody will ever

---

[66] Carey John, Dunlap William, and Pritchard, John R, (Eds) *International
Humanitarian Law: Origins*, (Leiden: Brill, 2003)

study this crime. As Italy clearly proves, justice is the servant of politics, and the law can easily be bent to serve whatever purpose if you are on the winning side.

Post WWII

Since the end of WWII there have been many wars involving ethnic cleansing, genocide, and every form or barbarity which can be classified under crimes against humanity and war crimes. There have also been many times of crimes against peace by going to war without any justified reason. The crime of war tends to fall at the feet of politicians and sometimes when they are unduly influenced by a belligerent military. Only a few years after WWII Britain and France started a brief war in Egypt over the well-known Suez crisis (1956) until America objected, and less than ten years later America entered Vietnam which many Americans rebelled against, but like Korea it involved barbarities by both sides. In this century the American invasion of Iraq (2003) could be and was seen by many to be a war of aggression, a crime against peace. Today the world watches aghast as Russia fights to occupy Ukraine, to name but a few.

Once conflict starts war crimes abound, when Lieutenant William Calley was convicted of the infamous Mỹ Lai massacre (1968) in Vietnam he was given a life sentence, but after public protests was placed under house-arrest and then released. This illustrated 'the selfishness of state, even those of liberal ones. We put our own citizens first—by an amazing degree'.[67] Morgenthau, a proponent for political realism argued that Lt Calley did what he was ordered to do, but unlike others did not escape attention, which was a deeply cynical approach given that Nuremberg established that illegal orders should not be obeyed. There were similar incidents of massacres by U.S. soldiers at No Gun Ri in Korea and Thanh Phong in Vietnam, often accompanied by excuses such as 'no one intended to commit them...but the politics of memory are precisely about contesting and confusing' that which is embarrassingly known to be true.[68]

---

[67] Kinsella, *The Morality*, p.399.
[68] Bartov, *Crimes*, p.xxiii.

It has been generally understood since WWII that carpet bombing is wrong when it is no longer a desperate necessity (and some may rightly argue that it is morally wrong *per se*) yet during the Vietnam conflict the use of napalm and airstrikes were widely used, and despite the use of smart bombs and target orientated missiles hundreds of thousands of non-combatants died in the Iraq war. Today, as these notes are written news has come through of the Russians targeting civilian areas in Ukraine. Film crews have verified what is a criminal act of aggression and a war crime against non-combatants; the chances of postwar justice are minimal.

Truth and reconciliation commission

Postwar justice which must have an underlying theme of rehabilitation, stability, and seeking peace has other possibilities of achieving these goals. Apartheid South Africa was a system of institutionalised racial segregation which existed from 1948 until early 1994. It was an authoritarian political culture in which the minority white dominated the majority of black Africans, Indians and others. Apartheid was known to be ruthless, protesters disappeared, imprisoned, and executed which led to a mild form of civil war. Every country has the problem of racism as it is a common failing of humankind based on the false assumption that skin colour is relevant. There was a form of petty apartheid which segregated public seating on transport and in parks, and the more serious which dictated housing and employment. It reflected the Nazi Nuremberg Race laws when the South African government prohibited inter-race marriage (1949) and even sexual relationships (1950). The Population Registration Act (1950) classified citizens into Black, White, Coloured, and Indian with further sub-clarifications. This appalling situation created both domestic and international protest and various embargoes on arms and trade were set in place and South Africa's system was frequently condemned in the United Nations. It was not surprising that at the domestic level the protests became more militant as did the government's response, leading to sectarian strife. Because of the external and internal pressure, the South African government made some reforms or concessions for Indian and coloured people, but understandably it did not appease the African National Congress (ANC) and the process to agreement was slow and at times violent. In 1990 the well-known leader Nelson Mandela, regarded by the racists as a terrorist, but to all others as a hero, was released from prison, and apartheid legislation was repealed on 17 June 1991 which led to the multiracial elections in April 1994.

The question was, as in any post conflict situation, how to resolve the problems of the past, and create a sense of peace and stability. The concept of the Truth and Reconciliation Commission (TRC) was authorised by Mandela (1996) and chaired by Archbishop Desmond Tutu. It amounted to a court-like body described as restorative justice, some of it given public hearings. Those who had suffered human rights abuse gave witness, and those who had perpetrated such deeds gave their testimony thereby requesting amnesty from criminal or civil prosecution. It was hoped it would cure the deep rifts of the past and lead to free democracy. As with any such event there have been critics as to its success, and many who thought it was a task well done. In 2000 the Institute for Justice and Reconciliation replaced the TRC.

One of the main purposes was to give a sense of national unity through the process of reconciliation. The TRC had three major committees one dealing with Human Rights Violations between 1960 and 1994, another the Reparation and Rehabilitation Committee not least intended to restore a victim's dignity and the third an Amnesty Committee for those who had applied. Those seeking amnesty had to prove their crimes were politically motivated, proportionate, and honest, and to avoid victor's justice all sides were not exempt from making an appearance.

Many have seen this effort for restorative justice as vastly superior to that of retributive justice such as the Nuremberg and post WWII trials. It certainly established the truth of the past, it could produce moments of reconciliation, and, it has been argued, brought South Africa out of its international pariah status, and improved the country economically. There were the usual problems of doubts, namely that the truth did not always emerge, that lies proliferated, and people did not want to be reminded of the past.

In any human endeavour there will always be the critics and supporters, but the concept of the Truth and Reconciliation Commission must stand out as a moral way to resolve a major catastrophe of a socio-political endemic evil. It is doubtful whether it could have worked after WWII at an international level such were the gross atrocities and millions of deaths involved. However, it may have worked as a healing process in single countries where tensions within a community continued for many years, not least over the French struggle with its Vichy problems and Italy divided on the values of fascism and who supported who. Above all the major value of the South African endeavour was that it underlined the possibility of a healing process because after any major or minor war the need for peace is paramount, and even if retribution and reparation are regarded as right for a time, reconciliation must be the final aim.

Final thoughts

For most people, especially those with liberal tendencies demand some form of a code of conduct in war as essential, because once the crime of war is unleashed the horrors and barbarities needs some form of restraint. Even non-liberal states need this if only to protect themselves and the soldiers who have become prisoners of the enemy. In the post war scenario, there remains the long-held tradition that a sound legal system does not permit political interference, but whether this is tenable at an international level is questionable, and the constant question of whether it is victor's justice cannot be ignored. The Nuremberg trial was a distinctive moment and assisted in preparing the foundations of stability in Europe. It was not resolved by castrating the enemy which Churchill in a moment's anger once suggested, nor by mass murder as Stalin had hinted and demonstrated at Katyń, or even his purge trials in the 1930s. The Western powers attempted a *bona fide* legal trial based on domestic law traditions, but it was pressure applied by the Americans, then backed by the British and French. The Russians complied but still wanted all the defendants executed. It is immediately apparent that such efforts as War Crimes Tribunals would find some impetus from liberal states, non-liberal states have never held such trials just settled their retribution in their own way.

Various trials have been held in the international courts with a varying degree of success, but only when possible. At the turn of the last century an attempt was made with what is called the Rome Statute (Agreed July 1998, in force July 2000) which was an attempt to give the ICC (International

Criminal Court) the right to determine its own jurisdiction. In part two, Article Five limited the ICC to the most serious crimes of genocide, crimes against humanity, War crimes and the crime of aggression. This may have raised hopes around the world, but the legal landscape and world politics are in a constant state of change. The hope was that evil can be restrained, but Pol Pot responsible for the Cambodian genocide could not be brought to justice, no more than it is likely that Putin can be taken to an international court for a war of aggression. In domestic law the system is supported by a police force, and outside the massive limitations of the United Nations this is impossible. The major problem is, as cynics might add, always political interference. Even in liberal states they will seldom put their own soldiers and resources at risk to bring a war criminal into the dock, and so for example trying to capture Pol Pot was too risky. There is also the inclination to take more seriously crimes against foreigners compared to one's own citizens, but despite the outrage that might be felt in places such as in Myanmar with the ethnic attacks and Cambodia's murders over ideology it remains in insoluble problem. There is also the perpetual fear of victor's justice, epitomised by and American Air Force general: '"I suppose if I had lost the war, I would have been tried as a war criminal", said Curtis LeMay, who targeted some sixty-three Japanese cities for annihilation by American bombing in WWII. "Fortunately, we were on the winning side."'[69]. The International Court of Law demands a military defeat first before it can even think of coming into action which is why Stalin and Mao Zedong never faced any form of justice. Nuremberg demonstrated the temptation was one of punishment, but the ideal policy was to reconstruct and establish peace. It is a highly delicate balancing act, and 'overheated moral judgments and particularly personal retribution, Henry Kissinger implied, risk undermining a peace'.[70] Any postwar war crime tribunal relies entirely on the liberal states who have the legal norms at hand, but who will always have the interests of their own citizens first. The moral and legal values postulated in liberal countries is irrelevant to many others, and organisations such as Amnesty International or Human Watch may have some impact in liberal countries, whereas it is meaningless to many others, including the superpowers of Russia and China, with this vast difference

---

[69] Kinsella, p.400.

[70] Kinsella, p.400

between liberal and non-liberal states making universal law or conduct seemingly impossible.

Back in WWII, Jodl wrote that 'the preservation of the state and people and the assurance of its historical future…give war its total character and ethical justification' if the word 'ethical' can be associated with this human conduct.[71] Telford Taylor, counterclaimed when he wrote 'it was high time that such antediluvian and essentially murderous paeans to the morality of war be buried'.[72] This comment is a constant reminder of the Nuremberg Trial where it was decided each person was responsible for their own actions, despite who holds the power. As primeval as Jodl's statement was it does not appear that humankind has travelled far, and it raises the question as to whether law is adequate to define the ethics and morality of war, but its universal acceptance remains unlikely. Where, in Western law the jury system is utilised, members of that jury are scrutinised to establish their neutrality, impartiality, integrity and still errors are later discovered, once national politics enters the international scene the dangers of a fair trial can be questionable.

Others argue, perhaps with some justification that only the political leaders who initiated the war should be put on trial, which was probably why Göring received so much time in the Nuremberg trial because he was regarded as the dead Hitler's stand-in. There is a case to be made that the person responsible for unleashing the dogs of war remains the central culprit but that has always been evasive. However, although it is a complex situation, politicians whether democratic, fascists heads, leading figures in an ideology or religion who start a war are the major guilty elements. Political bluster can be used to encourage support from the citizens and the military, and consequently the leaders are the ones who should be prosecuted. This could only happen to those who lost the war, not the victors even if guilty. Such is the hell and fury of war it leaves space for over-heated or misguided men to commit crimes in which only the rules of the jungle seem to work, and not all commanders can control what happens. This has led some to argue that it is unfair to punish men caught

---

[71] Taylor, *The Anatomy*, p.637.
[72] Taylor, *The Anatomy*, p.637.

up in war, but then there would no need to punish those who perpetrated such evil massacres as at Lidice, Oradour-sur-Glane, Katyń and endless other horrendous crimes, which goes against every normal instinct and norm.

The world has often been described as becoming smaller, describing the way modern technology of communication systems, travel, ballistic missiles, have made once faraway neighbours closer with all the economic, political, and military influences just next door. Progress has not always been beneficial and the sense of international anarchy of conflicting states, ideologies, sometimes religion, along with the habitual desire for wealth and power has left the world sitting on a precipice. It has become evident that international laws and institutions cannot change the situation as will be explored in the next chapter. Some so-called realists believe idealistic and legalistic arguments are useless, and the political realist will argue that the only salvation is the balance of power based on the fear of global destruction. The Nuremberg Trial set a precedent in many ways, not least the sense of stability that followed WWII but punctuated by the emerging Cold War and then a series of wars to the current day indicating that international law, despite the high hopes cannot prevent war. As many hoped the Great War was the war to end wars, others hoped that Nuremberg would give a period of peace and that the horrors of war Nuremberg trial had revealed, along with the knowledge of nuclear conflict, would be a deterrent against war.

St Augustine's theory that war should establish peace makes eminent sense, and hopeful as it seems it always appears to fail at the human level, as domestic laws need a police force to uphold the system because people daily break the law in every area. International law has in this postwar study failed to stop man's propensity for making war with its destructive and cruel nature. There are wars all around the globe many too remote to make headlines, many internal, some based on religious bigotry, some on human rights, others based on ethnic issues and economic suffering, and at the time of writing Russia has invaded Ukraine with ramifications not just in eastern Europe but across the globe. The only hopes in a postwar situation, if they exist, will be explored in the next chapter as it demands an international guardian, or a hope in some form of developing a common morality on critical issues, or even an international forum to study the aspirations of such experiments as the Truth and Reconciliation Commission produced by Nelson Mandela and Archbishop Tutu.

Chapter Nine - Law: 'Is or Ought'

Introduction

The words morality, law, international law, convention, all emerge in the vexed study of war including the term natural law. Throughout the history of humankind's self-examination of his propensity for war and how to avoid its destructive powers, scholars have used the term 'natural law' throughout recorded time. Colleagues, when prompted, offered various definitions and very definite thoughts on its values, for and against. Students may study the subject natural law in philosophy, theology, ethics, and jurisprudence, all holding the common feature of deep complex reading, often with bewildering inhouse vocabulary. This chapter will attempt to unravel some of the key issues and hopefully, where possible, in everyday language.

At its simplest natural law (*Lex Naturalis*) is based on observations of human nature and regarded as essential and intrinsic to humankind, and although related to various systems of morality and 'positive law' (that is law enacted by the state), natural law may be appealed to independently of state laws and even challenge an accepted morality code. Natural law implies that every person has inherent rights which are natural and can be discovered by reason which leads to such thinking as natural rights and justice. The Nazi laws (positive state laws) were often regarded as unjust because they contradicted natural law, and today (October 2022) in Iran there are protests over the way women are treated, over the state's law and having a morality police force. Many observers regarded the positive laws of Iran and their projected morality system as degenerate in the eyes of natural law, and this appears to be a widely held global sentiment. There are those who will claim that it is immoral to contradict the laws of the state or a long-held morality system within a community, but natural law can and often does challenge these views.

A brief history

Natural law was of early interest to the ancient Greeks, quickly taken up by the Christian faith, which often interpreted natural law as implanted by

God which brought forth criticisms. Others based it on pure reason, utilised by the French Revolution all of which 'resulted in a diminished respect for natural law during the nineteenth century. But natural law thinking has revived in the twentieth century, particularly since the Second World War'.[73]

Natural law has a long and turbulent past within recorded history and can be found in Aristotle who with other Greek thinkers often made a distinction between natural law and positive law. State law, it was justly argued, varied from place to place, city to city, whereas natural law was universal and applied everywhere. In his study *Rhetoric* Aristotle said state laws apply to particular people, but natural law was a form of common law applied everywhere. The Roman Marcus Cicero in his study *De Legibus* stated that everyday justice and good laws had their origin in what nature had bestowed on humanity because natural law, he implied, was serving humankind by giving a sense of common unity. This demand for a common unity has emerged again in modern jurisprudence debates. Cicero's argument acknowledged the existence of natural law and implied that it could be the foundation of positive law.

St Paul in Christian terms implied that even the gentiles had a sense of law stating that 'their conscience also bearing witness' to this fact.[74] For Christians it seemed that by acting to the guidance of natural law they were following Divine intentions. Aquinas added that any unjust law of the state is a perversion of law if it clashed with natural law. The Catholic (and later Protestant) Church propounded that humans could understand right from wrong because they had a Christian conscience. They added secondary explanatory precepts that drunkenness is wrong because it is bad for your health, and theft is wrong because it is bad for society. Islamic thinking has had different schools of thought on the subject, some Islamic scholars seeing it as survival of the fittest, but more often, as with Christianity regarding natural law simply as the laws of the Divine that humans can comprehend.

---

[73] Freeman M.D.A., *Lloyd's Introduction to Jurisprudence* (Sixth Edition) (London: Sweet & Maxwell, 1994) p.79.
[74] Epistle to the Romans 2: v15.

It is questionable whether man in his creation was wired with a conscience component as there was a time when it would have gone against a man's conscience to allow the children of an enemy to live, and for a cannibal it would go against any sense of conscience to let a stranger and his children go free thereby avoiding the tribal cooking pot. Nevertheless, this does not deny the discernment of a growing natural law which can be found through humanity's reasoning processes, as man has evolved through the ages. The sense of an inbuilt morality has grown as man has developed from his near animal existence in caves.

All religions have their own ethical systems and moral basis, they may differ in places, vary in emphasis, with some being changed as time passes and under varying circumstances. The Jewish faith established the decalogue, Islam has its code, as does Christianity, and one of the best ethical systems would be Buddhism. As humanity evolved through the various stages of development which involved living together natural law, helped by philosophical and religious thinkers, emerged as a powerful factor in the study of law and humankind's conduct.

By the time of the 17th century natural law was becoming prominent in English jurisprudence. The English Jurist Matthew Hale (1609-1676) who lived through the English Civil War and then the period of the Restoration claimed natural law was a gift of God so humanity could discern good from evil. He saw it as an antecedent to civil government stating that human law cannot ignore natural law. Thomas Hobbes (1588-1679) regarded natural law as directing how human beings survive. In his books *Leviathan* and *De Cive* he argued that natural law was founded by reason, and he produced 19 laws to back his theory. This amounted to an assault on the traditional views and was duly attacked by a Richard Cumberland (1631-1718) a philosopher and the Bishop of Peterborough. John Locke (1632-1704) also turned Hobbes' prescription around claiming that if the ruler went against natural law the people had a right to overthrow him. Locke associated natural law with Biblical ethics, but many have since argued that natural law was not the product of human reasoning but is spontaneous. It was also noteworthy that in the United States the Declaration of Independence proclaimed that the people of the U.S. assume 'the separate and equal station to which the Laws of Nature and of Nature's God entitle them'.

By the time contemporary jurisprudence was taking shape various channels of thinking about natural law had emerged. The different schools of thought persist, some argued that a constant repetition had become so widespread it was natural law, others that just laws are innate in nature, they can be discovered but not created, they can emerge from resolving conflicts, that they are God given, and law itself cannot be determined unless referenced to moral principles. It has been proposed that natural law was critical to the development of English common law and the tensions between the throne and parliament. To this day such tensions still exist between positive law and natural law in international law.

Modern concepts

Some positive law jurists want nothing to do with natural law, but others claim that between positive law and morality there must be some form of connection. It seems inconceivable to land up in a court of law as a defendant for trying to persuade someone not to commit suicide or for jumping into a private lake to stop a child drowning. This is based on the belief that a state law must reflect a morality or fair justice worked out by reason and not by a revelation. Legal history clearly indicates that law has made strides from executing starving people for stealing potatoes from a field or transporting them to the other side of the world. Few would dispute that positive laws of the state have been influenced by the general morality of the day and the ideals of social groups.

This does not mean that all legal systems must include a reference to morality or a sense of justice, but generally it is presupposed that state laws reflect an acceptable morality. It is known that in some countries the laws are immoral, such examples as Nazi Germany's race laws and South African Apartheid laws are but a few. When the British government decided that so-called illegal immigrants/refugees crossing the English Channel seeking help, would be flown to Rwanda (2022), this decision caused a justifiable outcry. It offended a sense of human rights, which many regarded as fundamental to natural law. This is a reminder of Aquinas who stated that man-made laws which clash with natural law are perverse. In most moral codes there will be some form of prohibition of the use of violence, a respect for promises, and many other aspects which help humans live together in some sense of harmony, which therefore indicate 'a core of indisputable truth in the doctrines of natural law', wrote a modern jurist.[75] For some positive law does not have to reflect any form of moral demand, but reflect a form of disciplined control based on the ideology by the political governing body legislating at the time. However, laws can be broken for bad or good reasons because man has free will. Natural law was criticised because of its theocratic view that laws are inbuilt by the creator into our nature, but natural law has not always been associated with a Divine lawgiver or human authority based on reason. Nor can it be associated with the laws of nature which are best expressed by such studies as the law of gravity, where the world of nature is more associated with noting regularities. Rather it is the belief that nature anticipates an appropriate end such as a seed in the ground eventually maturing into a life-giving tree, a growth which occurs naturally. This raises the question of the purpose of humankind, having an optimum state or end, which presumably means the development of a sound mind and character. These views are often associated with the teleological view (the end) which was Greek in origin, implying that what is critical is the 'end-product' to put it in its simplest terms. It has been argued that the proper end of human activity is survival which implies working together in harmony. However, this is the lowest element as there are, conceivably, greater ends for humankind, for some like Aristotle a greater intellect and for Aquinas a knowledge of God. Many thinkers are happy with the sheer survival factor because humankind cannot continue to exist without an association of individuals based on equity and reason. People wish to survive as there have been many examples of enduring suffering and misery to do so, which has more to do 'with social arrangements for continued existence not with those of a suicide club', the aim being to live together and improve.[76] This is an ideal which demands certain rules of conduct which must be recognised if such hopes are to be viable. These elementary demands recognise basic truths which are deemed the basis of natural law.

It is entirely rational that if survival is the core issue, then law and morals must have a specific content, because otherwise there is little hope. Without

---

[75] Hart H. L. A., *The Concept of Law* (Oxford: Clarendon Press, Second Edition, 1994) p.181.
[76] Hart, p.192.

acknowledging such rules humankind would have no reason to obey anything, and therefore reason must demonstrate the connection between natural facts and the legal and moral rules, which is why the appeal to natural law is considered by many thinkers to be the starting point. The most important being those rules which restrict the use of violence and killing, if 'Thou shalt not kill' was taken for granted people would be less vulnerable to one another. A basis of natural law accepts that all people are equal even though some are stronger than others, co-operation rather than domination which makes legal requirements necessary, it is a matter of this 'ought' to be the case rather than it 'is' a legal demand.

This disparity between powerful dominating people and the more vulnerable types comes to light in a more pronounced way in international law, with the vast disparities between states. This brings us back to Chapter One and the nature of man, seen by many as badly adjusted creatures who wish to exterminate one another, which clashes with the natural law premise of survival by living in harmony. 'But if men are not devils, neither are they angels; and somethings which makes a system of mutual forbearance both necessary and possible'.[77] Often the motive for obeying laws can be a mixture between prudence, with some people not interested in the welfare of others tending to put their own interests first. Because of Humankind's nature the reasons for good behaviour as prompted by natural law are often planted into positive law with its coercive system. There seems to be a great deal of sound sense in natural law forming a basis for positive law.

However, this does not resolve the question of what may be termed legal validity and moral values. There are still slave owning societies, apartheid systems, abuse of human rights and many other examples of where a state's law and moral system are sharply at odds with the foundations of natural law, especially where force is used to implement unfair laws. Often state laws will clash with the moral concepts of the same community, as with the recent outcry in Britain of sending refugees to Rwanda to save money and deter others, and the recent revolt in Iran (2022) protesting over the laws regarding the rights of women. The often-quoted belief that a legal system

---

[77] Hart, p.196.

must rest on a sense of moral obligation does not always hold firm. It can be true that morality systems influence law, the ending of capital punishment in Britain after centuries of executing criminals may be quoted in support of moral development, but often a community's morality can be flawed, especially if influenced by ideological or even religious bigotry. However, despite the potential flaws even the most ardent proponent of positive law is obliged to accept that a society's moral views impact on state law.

It has been suggested that in Germany following the atrocities of the Nazi regime that natural law was revived on the grounds that many of the violations of the corrupt regime had violated natural law which, as the ancient Greeks had claimed applied everywhere, and was in other words, international. It also underlined the fact that morally iniquitous laws cannot be regarded as law in its overall definition, namely that a law must be based on moral principles, which natural law demands. In the 1945-6 Nuremberg Trial it was considered to be wrong to obey orders which were illegal based on the premise they were immoral. The Nazis had like other regimes before and since used the law as one of the means to their ends. Natural law clearly indicates that a state law can be wrong and should not have the status of being called law.

International law

The term international law is associated with Jeremy Bentham (1748-1832) who defended it 'by saying it as simply analogous to municipal law' which is true of content but not form'.[78] The term International Law has been part of humankind's vocabulary for nearly two hundred years, but it has been suggested that the use of the word 'law' may be wrongly applied. International law lacks the secondary rules of change and adjudication which offers legislature and courts as in state law, and international law needs to be made explicit with credentials capable of inspection.

At the domestic level state law, positive law, is a matter of orders for the community and supported by police, courts, and punishment, but this

---

[78] Hart, p.237.

raises the vexed question as to how international law can be binding, and modern history has underlined this problem. It is generally assumed that the nature of law rests on the assumption that its existence makes some conduct a matter of obligation. The United Nations may make laws, but they are always prone to be blocked by the veto, and the law only exists on paper. As such the frequent criticism is made that it is only a talking shop. Most people assume law is a body of orders supported by threats. A law, it could be argued needs this support because in a modern state if there were no punishment of crime then robberies and even murders could soon become out of control. At the international law level there is no serious comparison with state law, as it cannot offer a police force, only limited punishment in the possible form of sanctions, cannot be internationally binding and for many critics the word law is undeserving.

Another issue with international law is that every state has its own sovereignty and that is such a dominating factor it can put international agreements into the background. The very word sovereignty it is associated with a leading person such as a monarch who is above the law, recalling the infamous statement 'L'étate c'est moi'. Even the word state can be vague if not tenuous in places, but it is more recognisable and acceptable than international law for many people. Most of today's states are controlled by their own domestic laws but not felt bound by laws outside their own boundaries. The very word sovereign carries the implication of independence and self-autonomy.

It could be argued that sovereign states can volunteer to subscribe to an international agreement, a sort of auto-limitation exemplified in treaties, pacts, and so forth, but history has clearly demonstrated failure here. Most of the major countries signed the Kellogg-Briand Pact (1928) to renounce war as a solution, but within three years this proved ineffective when in 1931 Japan invaded Manchuria.[79] There were no plans for enforcement and the legal terms were lacking definition. There was nothing to explain how

---

[79] Known as the Mukden Incident the Japanese flew a false flag as a pretext for the invasion. China appealed to the League of Nations followed by discussion. The USA was not a member which did not help, and a face-saving committee under Lord Lytton reported on the incident and by 1932 found Japan to be the aggressor who responded by walking out of the League the following year.

states could regard such an agreement as binding or be self-imposed, such is the nature of the sovereign state. These international treaties and pacts are best seen as signs of intent, a declaration of proposed conduct. However, a breach of such promises which can be cast aside when deemed necessary (as in domestic life) hardly makes them law in the sense of the domestic application of law. There is little historical evidence that any state feels bound by international self-imposed obligations, and this remains evident to this day.

States change their governments or change ideologies, new states emerge (Israel 1948, Iraq 1932) and previous international agreements cannot necessarily apply. Many of the international agreements are based on moral motivations and can be cast aside on the grounds that they are 'only moral' and therefore no obligation exists. Germany signed the Kellogg-Briand Pact, but Hitler emerged in 1933, a leader noted for being amoral seeing the pact as a mere moral formulation. China may claim a moral right over occupying Taiwan (based on a form of irredentist nationalism) as Russia may claim a right to invade Ukraine because of its history and the perceived threat of an encroaching NATO. Moral obligations can be argued against by a different viewpoint on morality, especially when not backed by an enforceable law as in a municipal state. Domestic law, and not morality puts a defendant in the dock, produces witnesses, evidence, and punishes the wrong doer, but international law is mainly moral and does not have a universally recognised legislature. If an international law is broken it may lead to protests, even sanctions as applied to Mussolini (1936) and the South African Apartheid Government (1948) but with the veto and without the force of domestic law with its threat of direct punishment international law is limited to telling off the offender. International law simply does not have the force or power of domestic law, lacking what may be termed a rule of recognition.

The ideal of the United Nations was to bring global communities together, to agree on conduct enforced by international law is a high moral aspiration, born from the failure of the League of Nations and the total disaster of WWII, with countries seeking a peaceful future. The traditional arguments of natural law of living in harmony, human rights, and peaceful co-existence were seen by many as the foundation of this international enterprise. Politicians, Jurists, and all forms of disciplines like many observers have serious doubts about how effective a body like the United Nations and international law can be effective. The United Nations can offer hope but, it is hamstrung by the veto, and by having no machinery to enforce its laws.

Some final observations

Humankind has evolved in many ways, not least in behaviour or moral patterns, from cannibalism, slavery, murdering neighbours, then through centuries of thought and reasoning to developing systems of moral behaviour and using law to live in some form of harmony. Morality has changed as the centuries have passed, burning heretics at the stake, torturing those who criticised the monarch, the Spanish Inquisition now all mainly belong in history books. Morality is not uniform and both in the past and today can differ from country to country, sometimes related to the general religious faith of a state. Human rights for male and female even today differ in various countries, in some places homosexuality is a capital offence, and although humanity's moral thinking has improved and different structures have much in common, a single morality system cannot be claimed to have a universal status.

The law of a country tends to follow the moral precepts of its community, in Britain, when a jury refused to find guilty a man who was unquestionably guilty of stealing a sheep, they did so as they were revolted by the fact that such a crime carried a death sentence, and the law was changed. It was for the reason of moral objection that capital punishment stopped in the UK and people have the right of protest, indicating that citizens can react to a law they consider to be morally wrong. Other countries may have a different moral basis where people outside the country think their neighbours have perverse laws. A country which dislikes another race of people, wants to punish homosexuality, thinks women are less important than men, may have public support based on their version of morality. Most people drive petrol or diesel fuelled cars and it may be in a few years they will be condemned on moral grounds as once drink-driving was not a legal offence but is now illegal and regarded as immoral and as times change so must the moral ethos.

The positive-state law must be obeyed because it is an order which is watched over by the police, there are courts to try them, and an ensuing punishment. In some countries breaking the moral law is also punishable, abortion, adultery are just two examples where the moral code has been subsumed by state law.

As noted above international law is based on moral standards for the harmony of nations, but it has no global jurisdiction, suffers from the infamous veto, and unlike the sovereign state cannot enforce a broken law except by expressions of disapproval. Sovereign states feel free to leave the United Nations or not join, may break treaties and promises, ignore the pressure points, and pursue their own direction. The United Nations is first class in terms of humanitarian crisis such as famine, floods and climate change, but will always have serious problems with political issues, ideologies and even religious faith.

There is no answer which is why natural law resurfaced after WWII. It has been cogently argued that natural law provides what may be called a universally felt common-sense morality. There are some fundamental principles which must be acceptable to most normally adjusted people. That it is wrong to murder must be a universal truth accepted by everyone albeit ignored by vicious criminals. This assumes that starting an aggressive war for greed, power, or political domination is immoral and wrong. Human rights relating to personal freedom as viewed by natural law condemn slavery or unfair imprisonment and support the right to live. Skin colour or physical features are not to be abused because they too fall within human rights, and along with countless other aspects of life.

In the ideal Utopian world, a gathering of nations should list the rights of people under this common-sense natural law and agree such conventions should never be broken. If they feel natural law is unreal, then perhaps it should be where all moral systems try to reach a point of agreement, a sort of reflection of natural law. This was part of the ideal of the United Nations, based on the experience of WWII, when the sheer survival factor demands humankind can only persist and enjoy life when individuals and nations are based on a sense of equity and reason and can thereby live in harmony. If not, the alternative is as the jurist H. L. Hart described as a suicide club.

It is widely accepted today that aggressive war is both morally wrong and illegal. But since WWII more people have died in war zones than during both world wars, and conflict is still ongoing, not least in the Ukraine, the world is also constantly alert to potential conflict between China and Taiwan with its possible ramifications, and rightly nervous about North Korea's playing with missiles dangerously close to its neighbour's borders and over Japan. Unless the United Nations has something like a state system of policing backed by a court system it can never work in terms of stopping war, especially when superpowers are involved. There is no doubt that the United Nations has high ideals, has done some exemplary work in trying to resolve conflict in some individual countries, it also provides a platform for moral opinions, assisting in areas needing assistance, and making sure the world is aware of the dangers. The sadness is that it is powerless to stop major wars which may lead to global conflict and the danger of humankind's self-annihilation of mutually assured destruction (MAD).

The issue of moral systems and/or natural law remains confusing. It relates to the question about the use and understanding of 'ought' once described as the major moral verb when used correctly. A bricklayer may say 'we ought to build this wall first' implying that it is the right thing to do, but it could be based on the intention of simply impressing the owner of the land, but it is not an obligation. Professor Campbell wrote that 'For whatever else the moral consciousness may be, it is at least a consciousness of oughtness or obligation'.[80] When a person is faced with a moral question such as should we kill this prisoner and another replies 'we ought not' there is the distinct sense of a moral obligation to do the right thing. Campbell furthered his analysis by stating 'I feel free to assume, the moral consciousness is at least a consciousness of obligation.'[81] When a person uses 'ought' in a moral situation the implication is that some actions are based on a moral sense of obligation. He concludes his thinking with 'the existence in man of a moral consciousness—a consciousness of unconditional obligation, of a categorical imperative—I propose now to be

---

[80] Campbell C A. (Professor of Logic and Rhetoric) On Selfhood and Godhood (London: George Allen and Unwin, 1956) p.181.
[81] Campbell, p.181.

established'.[82] The implication being that it is not a subjective viewpoint but an inbuilt law of obligation. It was the German philosopher Immanuel Kant (1724-1804) who first wrote of the categorical imperative as an unconditional moral obligation which is binding in all circumstances and is not dependant on a person's inclination or purpose. Without becoming too wrapped up in philosophical reasonings most people would agree, often utilising the word conscience that we feel inside ourselves moral obligations.

This book has explored aspects of the morality of war, because morality touches all our lives, effects the individual and corporate conscience, but such is war once it starts neither morality, natural law, nor positive law can resist the meaning in the phrase 'that the end justifies the means'. Pope John Paul II in addressing a group of soldiers described human society at peace as a Utopia, and he was uncomfortably close to the truth. In the event of war only the laws and rules of the jungle apply, which is a sad conclusion based on humankind failure.

---

[82] Campbell, p.202-3.

Chapter 10 - Today

Introduction

During the 20th century the world has witnessed the growth of superpowers and the decline of Europe as a military power. During two world wars followed by a series of proxy wars and other conflicts Europe became a tinderbox for conflict and Ukraine and 'Putin's War' against Ukraine has added to this anxiety. There is a necessary demand for reform, not least because a war involving superpowers and their alliances may well lead to annihilation of life with the dangers of WMD. One fifth of the way through the current century there is potential for such a catastrophe, and it comes at the same time when humankind's attention should be wholeheartedly concentrated on averting the crisis of climate change. It is a universal and extremely dangerous crisis, yet humanity's preoccupation with political dominance invoking the dangers of conflict still envelope our thinking processes. At the time of writing (2022) Marx's warning that tensions are created by global market economies remains all too pertinent. The sanctions applied against Russia to express disapproval of the invasion of the Ukraine have proved to be a double-edged sword. Many parts of Europe were heavily dependent upon Russian gas and oil, and many parts of the world are suffering from a rapid economic decline. The world remains a politically divided community with the once powerful communist Russia becoming more like a totalitarian state, other countries following similar patterns, the liberal democracies reeling between political swings from the left to the right, and in places what has been called kleptocracy, where corrupt leaders use political power to expropriate the wealth of the people and the territory they govern. Capitalism demands a stable society in which to flourish, but xenophobia, nationalism, authoritarianism continues to grow, and religious demands have re-surfaced with a vengeance with extreme Islamic groups developing a two-way visceral hatred with the persistence of terrorism. Despite the warnings of the death and carnage of the 20th century bellicosity remains a characteristic of humankind, and the superpowers with their associated political alliances continue to build and sell more sophisticated weapons devised or invented by clever scientists, all technically non-combatants.

Much has been written in the past and today about non-combatants and their human rights. There is one issue which past writers have raised when they state that only the sovereign leader may declare war, it cannot be the initiative of another person or groups. However, there are groups of non-combatants who can lead a society to an unnecessary war or prompt one by the nature of their work. The first are politicians who as individual and groups can choose the wrong path, then those who sell armaments and scientists who develop them.

Non-combatants: Politicians, arms traders, scientists

Politicians always seek popularity to retain their sense of power. In democratic countries it is done through voting, political parties trying to find any form of charismatic type leader with elections mainly focused on promises of a better economy and more wealth, rarely on foreign policy and relationships with other countries. In non-democratic areas it is also based on popularity and charisma, with pictures of a younger President Putin with his shirt off and fishing or riding a horse in some wild background and playing ice-hockey. In some it is based on being the top authoritative person of an ideology or religion. The clue to success is for leaders to please their people, this can be done through economic welfare, social care, housing, health, and education which all makes good sense. This explains why in most general elections economic promises proliferate. However, there is another more sinister underlying theme based on making promises of a sound defensive system, not unreasonable, but which can soon degenerate into bellicose language about perceived enemies, the need for national pride, flying the flag, speeches about the mother or fatherland, our proud national history, heroes of the past inferring the current politician would like to be one. In democratic and other types of societies it is nearly always the political leaders who somehow prompt war. As in all walks of life there are angels and devils in any group of people. There have been outstanding political leaders, some born to the job as the saying goes, some mediocre and less effective, but the danger are the bad ones seeking glory and a place in history. They will be returned to power in a democratic system if they can evoke a bellicose response from the population who can sometimes be too easily stirred with jingoistic language.

In the capitalistic world economics often governs the minds of politicians, and the manufacture of arms and their enhancement by advancing technology makes for greater sales and national income as well as defence. A sudden build-up of arms may be seen as beneficial, but it can backfire. Sir Edward Grey wrote:

> 'Great Armaments lead inevitably to war. The increase of armaments…produces a consciousness of the strength of other nations and a sense of fear. Fear begets suspicion and distrust and evil imaginings off all sorts, till each government feels it would be criminal and a betrayal of his country not to take every precaution, while every Government regards the precautions of every other Government as evidence of hostile intent'.[83]

To which Howard raised the question as to whether 'there was any answer to the witches' brew which imperialists, finance-capitalists, militarist, and armaments-manufacturers were cooking up for mankind'.[84] International Sales of sophisticated as well as normal weapons (guns and so forth) have become a crucial part of international life, and can often be accused of causing havoc in some regions where instability is well known, and even lead to an increase in some regional powers, with distinct possibilities of nuclear proliferation. Arms in themselves do not necessarily lead to war yet no one doubts they can exacerbate problems.

Finding information about the arms trade is no easy task, there have been various bodies or institutions which provide some curious information, the most significant being SIPRI (an independent international institute dedicated to research into conflict, armaments, arms control, and disarmament established in 1966). In one of their charts, they ranked arms sales by nations during the period 1950-2019 with the U.S.A., Russia, the

---

[83] From Grey of Fakkoden, *Twenty-Five Years* (London: Hodder and Stoughton) 1926) Vol. 1, p.91.

[84] Howard Michael, *War and the Liberal Conscience* (London: Hurst, 2008) p.55.

U.K., France being in the top four, followed by Germany, China, Italy, Czech Republic, Netherlands, and Israel. By 2020 the lead recipients were India, Saudi Arabia, followed by Australia, South Korea, Egypt, China, Qatar, U.K., Pakistan, and Japan. The weapons can be everyday weapons used by police and foot-soldiers, to highly sophisticated weapons including missiles, planes and naval equipment. Over the past few years with cyber-attacks becoming prominent there has been a vast amount of money spent on cyber-security software which has proved critical for defence systems. In terms of world trade this all amounts to billions if not trillions of pounds.

The U.S.A. remains the largest exporter of weapons, and during President Trump's time in office the portfolio of customers, it was rumoured, grew significantly riskier. Trump showed little interest in the UN effort to grapple with the danger of small arms sales. It has also been noted that the British have often approved sales to countries subject to such trade restrictions. An Arms Sales Risk Index was created a few years ago to measure the risks of selling to some countries, but perhaps Marx was sadly all too correct when he said money is the driving force of history. There are some countries known for human rights abuses such as Bahrain, Bangladesh, Colombia, Egypt, and Saudi Arabia but not subject to any sanctions for such sales.

In the sale of arms countries use an EUC document (End-user certificate) which certifies the country which will be using the purchased weapon, but the system is far from safe as it is well known that a sovereign state often feels free to renege on agreements, it may be for financial profit, or even a secret agreement, and often because they feel they have a right to sell on goods they have replaced. This may well result in a small rogue nation obtaining weapons which are out of date in the major league but can have a devastating impact in the local area. It can also lead to the ironical situation that the country which provided the weapons in the first place may find itself in conflict against its own weapons.

In Yemen probably because of the Arab Spring democratic protestors took to the streets to overthrow the 33-year long rule of Ali Abdullah Saleh and the conflict intensified about 2015 with Saudi Arabia becoming involved. It resulted in a major humanitarian crisis and although figures remain vague it is generally estimated that some 250,000 people have been displaced and well over a 100,000 people have been killed. In May 2019 the American State Secretary Mike Pompeo as an emergency issue demanded the need for $8 billion of arms sales to Gulf Allies such as Saudi Arabia, the United Arab Emirates despite opposition from Congress, and Pompeo was criticised for violating the Arms Export Control Act. Britain, who had sold arms to Saudi Arabia also received criticism for playing a dual role as diplomatic penholders for the Yemen at the UN and also military advisers to the Saudi coalition based in Riyadh. For some critics, political principle and money have become too closely merged, because the war causing such pain and suffering is simply immoral and wrong, but it is always assisted by selling weapons.

Either way there is no doubt that the weapons trade makes considerable money in terms of trade, but it is equally dangerous for world peace because it is evident that weapons are not always used for defensive reasons. In an ideal world there would be no need for this trade, but the world is far from Utopia, but the tightest controls need to be exercised and not weakened by the desire for money.

As politicians and arms dealers/manufacturers are technically non-combatants, but possible targets for attack, the same applies to scientists who are involved in devising new weapons as technology develops. As noted in Chapter Seven it was the fear of the German Nazi regime developing a nuclear weapon which prompted politicians to ask sometimes reluctant scientists to produce the first atomic bombs. However, while soldiers operate under given rules of conduct, especially not obeying illegal orders, by the same token scientists should not work on or produce illegal weapons such as gas and bio-chemical weapons. This would be accepted by most, but the constant fear that the enemy may be busy producing such weapons makes it for some a necessity of war, thus many such laboratories remain secret or veiled under some other pretence of work.

There is little doubt that in this non-Utopian world humankind has created there is a need to create weapons if only for the need of defence, but this raises the question whether defensive nuclear weapons or any weapon of mass destruction used even in retaliation is justified given that it could involve the annihilation of entire life on the planet. There have been efforts to limit the proliferation of nuclear weapons to this end, and a careful watch when states like Iran are suspected of building their own on grounds of sovereignty. Every wealthy country has highly secret war development centres where new ideas are studied in anticipation of producing better, and therefore more destructive weapons than the presumed enemy.

There is no doubt in this writer's mind that some scientists may think of a new weapon but decide to take the idea to the grave with them. Others may be less morally inclined. For argument's sake let us suppose a group of scientists come up with an idea that they can eliminate an entire race of people by secreting a dangerous genetic code into food supplies which will be dormant for a time not to alert suspicion, and then an entire race of people will be eliminated. Such a weapon conceived by an individual scientist would not only be illegal and totally immoral but if conceived under the auspices of a government that country would be guilty of planning genocide.

Politicians and state leaders are the critical people in this scenario, but the monetary benefit of armament manufactures and traders, as well as misguided scientists are all capable of bringing on war. As in all groups of people there are angels and devils, but the devil element will flourish if nations fail to live in harmony or honestly accept some form of international agreed control.

There Will Always Be War

This book started with trying to understand the nature of humankind with its propensity for conflict and war, an issue which has been a human characteristic throughout recorded history and has not changed one iota. It dwelt briefly on man's sense of corporate identity reflected today through a sense of national identity based on memory, and then explored that from the earliest of times some of the greatest world intellectuals have tried to understand humanity's thinking processes why war starts and when it is wrong. It appears that those cynics who regard war as natural as the plague or floods are probably closer to the truth than idealists. It therefore seemed pertinent to study the high ideal of pacifism which cannot be achieved unless there is total global agreement which is unrealistic, and this was followed by a study of positive realism based on the theory that the only thing which counts is political pragmatism to maintain the balance of power, A school of thought with a long history which was examined, found to be an ongoing feature of modern thinking and somewhat cynical.

Based on theories of St Augustine the modern world based its attitude towards wars of aggression as wrong, but wars of defence as just. Humankind's character naturally smudges these clearcut parameters with justifying pre-emptive attacks, but with many suspiciously grey areas. Less justified are wars of prevention when the causes for aggressive war can be dressed up as defensive, as wars of prevention are based on speculation and sometimes misguided assumptions. Another grey area are wars of intervention, sometimes justified if based on humanitarian grounds, but often used as an excuse for aggression. The current era of modern terrorism has increased the need for defence, but in recalling that some forms of terrorism demand more attention as the resolution may be found in trying to correct human abuses which cause some terrorism. There is a difference between a terrorism provoked by a serious injustice, to that which wants to convert people to a religion or a way of life.

More defined are the rules or conventions regarding conduct in war, involving such major issues as prisoners of war, surrender, torture, suicide missions, reprisals, hostages, sieges, blockades and even sanctions. It also involves the vexed question of the innocent non-combatants. As always, the rules are frequently broken on the battlefield and at command level on excuses of military or political necessity often utilised to justify criminal behaviour. The postwar situation is examined especially relating to victor's justice and the need for trials and punishment to establish peace and stability, which in the light of recent history also appears to have failed despite some genuine efforts.

It is clearly the case that humanity is incapable of self-control or mutual agreement. This was explored in a brief chapter on the ways international law and global bodies like the League of Nations and United Nations simply fail at the dangerous crisis level of war. Morality systems while often having much in common may differ, especially where they are attached to a religious faith, which leaves what is known as natural law which seeks commonality of moral belief on a global level. This would mean establishing a form of moral absolutism demanding that some rules can never be ignored, which means war should no longer happen and living in harmony would be secured. Sadly, this is not within humankind's make-up, because there will always be rule breaking, which is seen in nations persisting waging aggressive wars, and in the conduct of war when the oldest war conventions are frequently broken. Even in the nuclear age conventions, laws, treaties can be broken in the face of imminent catastrophe, and it can only be hoped that the moral urgency eventually prevails. In medieval Europe the Church was often able to be the measuring rod of behaviour and at time there were hopes that it could be the key to a global community and peaceful harmony. It failed as it was too often politically motivated, abusing its power for self-seeking purposes and at times simply unjust. This was not the fault of the Christian faith, but the church was run and organised by men who had all the failings of the rest of humankind, greed for wealth, power, influence and dominance. The Russian Orthodox church has a section on *War and Peace* in its major work the *Basis of the Social Concept of the Russian orthodox Church*. It clearly distinguishes between an aggressive war which it condemns and that of a justified war and which basically follows the Christian theme of a Just War. It was therefore something of a surprise when the Russian Orthodox Church seemed to support President Putin's attack in Ukraine, once more underlining the fact that clergy do not always reflect the Christian demand of loving one's neighbour.

Times change as do the contexts, and today the difference between combatant and non-combatant is totally inconsequential. A nuclear weapon is the apex of the bomb and as in carpet bombing everyone becomes a target for death but on a gigantic scale with the prospect of more death caused by fallout. It appears to be a fact that international law and even agreed moral principles will fail unless there is a form of agreement which would abolish war. Therefore, nations must continue to labour at moral restraints and not lose sight of fundamental rights to live, which continues to appear to be a forlorn hope. It is generally agreed that genocidal wars, or conflict based on ethnicity or religion are fundamentally wrong. This is equally regarded true in the use of biochemical and gas weapons, weapons which rule out any opportunity of applying the arguments of proportionality, and the only hope if peace is needed is some form of reconciliation which cannot be delayed. Sadly, although mankind has made huge steps scientifically, medically, technically, our failure to grasp the lessons of history will affect the future. It was the German philosopher Georg Hegel who once said, 'the only thing that we learn from history is that we learn nothing from history'.

A medical doctor friend of this writer who read a book on the Nuremberg Trial admitted that his reading took him beyond the war films and glories of victory to those who started a war. He was amazed how ordinary people were convinced to carry out orders which resulted in an industrial scaled atrocious barbarity, which even when we were tribesmen nothing so wicked happened, but it occurred in the so-called developed civilised world. It took a history book to open the eyes of a medical professional as to humankind's propensity for killing one another, which poses two questions, the first is whether historians have failed or whether history has been neglected. Secondly, if John Doe takes an interest in world news, not just the problems of inflation, employment, or some scandal, then the ongoing problem of self-destruction which humanity increases can be seen as still far too active. The sadness is that those who claimed war cannot be avoided appear to be correct according to history.

Today's world

Tragically during the 21st century there have been and continue to be a series of armed conflicts which hardly make the news headlines. On the other hand, the Russian attack on Ukraine (2022) has dominated the world's media because of the serious possible global ramifications which will be explored later. Wars come in various forms, international conflict, localised, civil wars, or involving many as in the world wars. Civil wars can also be divided between those based on political motives, ethnic hatred, focused on religious divisions within a faith or a clash of different religions. In Mexico there has been an ongoing civil war between the government and powerful drug cartels, a drug war which has killed thousands of people. There is the development of terrorist insurgency which can start and finish in one country or be an international problem.

The list is an endless catalogue of horrific war zones which only makes the news headlines from time to time. Myanmar (the old Burma) has experienced internal war for years and is now controlled by a dubious military junta. In the Sudan a civil war since 2013 during which UN peacekeepers failed to stop nearly a million and half people from being displaced and with a rising death toll. There has always been UN troops in and out of the Republic of Congo with some half a million refugees fleeing for safety. Afghanistan has been in conflict as the Western occupying powers tried to oust the Taliban, but under American orders promptly left with the Taliban immediately reasserting their authority (August 2021). In Iraq the war of 2003 but American troops stayed longer to counter the Islamic State group (IS) and there was a bloody war in Syria (2011) which gave rise to ISIS involving major world countries including Russia and France. In Yemen, mentioned already there is civil war with Saudi Arabia involved. There appears to be perpetual conflict between India and Pakistan over Kashmir, and perpetual conflict between Israel and its Arab neighbours which has been ongoing since 1948. The heat of the conflicts rises and falls but do not seem to be resolved. This would be true in Somalia, in Darfur, (a region in western Sudan), and Peru with separatist armies in constant conflict, West Papua (Western new Guinea), Angola, Mozambique, Nigeria fighting the Boko Haram terrorists, and many more countries where dangerous instability may suddenly reach boiling point and internal war. These conflicts reflect hostility between different groups of people based on ethnicity, religion, politics, and even cultural lifestyles. They reflect the usual reasons for war, territory, dominance, wealth, political influence, or authoritarian power, and often destroying the essential infrastructure which takes years to rebuild, to mention nothing of the loss of innocent lives. Any conflict which produced a thousand deaths is a war in normal terms.

The anticipation of war creates many problems, and the international community often looks with a sense of trepidation at what are often called rogue states. Various treaties have been signed to stop the proliferation of nuclear weapons, but once again sovereign rights override many international agreements. In Iran, currently undergoing harsh press reports on human rights regarding the treatment of prisoners and their own women, led to an astonished global audience watching the Iranian football team in the 2022 world cup hardly singing their national anthem. The Western powers suspect Iran of supporting terrorism, and there is the constant fear that their research and work on nuclear power is not just for energy resources.

In North Korea, sharing borders with China and Russia and south Korea, has been led by the current leader Kim Jong-un (from 2011) who is now the third member in the family dynasty, shown for the first time on television with his daughter (November 2022) with some political commentators noting he was probably introducing the fourth dynastic leader.[85] The ongoing problem has been their development of nuclear weapons and missiles to deliver this danger, constantly making the world alert to a potential threat. In 2017 there was a moment's crisis with much political rhetoric between President Trump and Kim Jong-un which quietened down but is now heating up again. In March 2022 North Korea conducted a successful ICBM launch for the first time since 2017 and six months later passed a law declaring North Korea a nuclear state. Kim Jong-un is often in the headlines watching or announcing missile trials landing some close to the South Korean coastline, others launched over Japan, and constantly warning America that his ICBMs can reach the American continent. It is a case of serious bellicosity, but the world hopes that it is only a small country with a leader trying to make himself appear a world figure and he would not dare start a war. The danger is that a misjudged political move, a miscalculation, or a technical error may trigger a potentially dangerous war. How to restrain such a figure as Kim Jong-un is beyond the capability of the UN and raises the question as to whether China or Russia have any control or does not mind having a naughty boy in their backyards. It has been suggested on some documentary media channels, including the BBC, that both China and Russia are content to have North Korea with such weaponry as their buffer state.[86] More significant than North Korea are the dangers of a clash between China and America over an offshore island.

---

[85] BBC World News, *Unspun World* by John Simpson, 26, November 2022.
[86] Ibid.

Taiwan

What does grip the news headlines are the potential conflicts which could involve the nuclear superpowers. At the time of writing there is deep concern over China's wish to control Taiwan based on the irredentist nationalistic claim that the island, once Formosa which was a short-lived republic in 1895, traditionally belonging to the Chinese mainland, but retaining the name Formosa for a time. The Japanese occupied it during WWII, and in 1945 the ROC (Republic of China) forces with assistance of the Americans landed to accept the Japanese surrender. After the war the civil war in China between the Chinese Nationalists (Kuomintang) led by Chiang Kai-Shek and the Chinese Communist Party led by Mao Zedong raged unabated, and by 1949 the Chiang Kai-Shek forces were compelled to retreat to Taiwan where the ROC has been ever since. They had other smaller islands which were eventually taken by the Chinese communists, but the ROC retained Taiwan and still claimed right of authority on the China mainland which was somewhat hopeless and senseless. In today's world the tensions remain with China making loud threats to retake what they believe to be their territory, not just threats but imposing their presence at sea, which has caused tensions between China and Japan. The Americans during President Biden's time have made it clear they will help Taiwan in the event of an attack.

As Taiwan is only 110 miles away from China's coastline and with China's President Xi Jinping making it clear that Taiwan would be 're-taken' the tension in the China Seas is almost palpable. When Biden first stated that America would support Taiwan many took his words as somewhat casual and without too much significance, but he repeated the promise several times, making the relationship between China and America uncomfortably frosty. When the House Speaker Nancy Pelosi planned a trip to Taiwan the Biden administration tried to dissuade her but failed which initiated another sharp decline in Chinese relationships, with China sending its warplanes into Taiwan's defence zones as an act of warning. This tension continually attracts the world press because the slightest miscalculation or misjudgement, even the wrong words could precipitate conflict between two superpowers with the fear of total annihilation always lurking in the background. The question arises again as to the inability of international law and the United Nations being able to have any form of control on such a situation. The only hope is that some form of natural law will send out the message that such potential destruction is wrong and every resource apart from weapons and military threats are examined and acted on with care.

Ukraine

Just as worrying as Taiwan is the Russian invasion of Ukraine which in the Western countries, we only hear our own media reporting, and for a more objective view it is important to understand something of Ukraine's turbulent history. Ukraine is the second largest European country (Russia being the largest) and covers some 230,000 square miles with a population of over 40 million. It has borders with Russia, Belarus, Poland, Slovakia, Hungary, Romania, and Moldova, which may explain much of its difficult history. The capital is Kiev, and the official language is Ukrainian, but many are fluent in Russian. Ukraine was and is a gigantic farming area which was slow to industrialise and despite the discovery of coal in the Donbas region remained mainly agricultural.

Ukraine was ruled by a variety of other powers for some 600 years including the Polish-Lithuanian pact, the Austrian Empire, the Ottomans, the Tsars of Russia, at one time partitioned between Tsarist Russia and Poland. During the Great War (1914-18) there was fighting on Ukrainian land with the Ukrainians divided between Austria-Hungary and the Russian Imperial Army. The fighting continued in Ukraine caught up in political instability and military commitment with the left-wing inclined Ukrainian People's Republic, basically centred in the west of the country which the Poles annexed for a time. Following the 1917 Russian Revolution a Ukrainian national movement emerged to become the Ukrainian People's Republic but this was short lived because the Bolsheviks forced Ukraine into becoming the Ukrainian Soviet Socialist Republic (1922). The eastern part of Ukraine was hit by a major famine in 1921 mainly caused by instability and war.

Stalin in the 1930s was held responsible for what has been called the Holodomor which has been described as a man-made famine. The number of Ukrainians who died as a result is impossible to ascertain, but it is generally accepted that three to four million citizens died. The reasons for this famine are equally uncertain. Some believe it was Stalin's way to eliminate the simmering independence movement, others that it was a result of rapid Soviet industrialisation and the badly organised agricultural collectivisation when Ukraine was subjected to unrealistic quotas. As recently as 2022 Pope Francis on the anniversary of the Holodomor (observed on 4 November) compared the tragedy to the current invasion of Ukraine.

The war of other nations soon embroiled Ukraine again when German armies invaded (22 June 1941) with ferocious fighting around Kiev. Hitler recognised the Ukraine as the bread or wheat-basket of Europe because of its potential massive agrarian output. Ukraine has extensive fertile land and in more recent times became the largest grain exporter in the world, with many countries dependent on their exports. Many Ukrainians hoped the Germans would be their liberators from Stalin's grip, but more were horrified that the Nazi agricultural policy was no better than Stalin's and watched in terror as they saw first-hand the genocidal Nazi policies happen on their doorsteps. Many Ukrainians fought alongside Russians, some chose the Germans, and again an insurgent nationalistic army materialised known as the OUN (Organisation of Ukrainian Nationalists) seeking independence, who sometimes allied themselves with the Nazis, and became involved with horrific massacres of Poles to try and create a homogeneous Ukrainian state. Ascertaining the figures involved is an impossible task, but over four million fought for the Soviets, and it is believed that during WWII over six million Ukrainians were killed including over a million Jews. It has been estimated that of the 8.6 million Soviet troops lost, 1.4 million were Ukrainians.

For most of its history Ukraine has been governed by others, occupied, partitioned and under foreign influence, a country divided by ethnic or language barriers between the Ukrainian and Russian speakers. Nevertheless, constantly seeking independence but with others whose first language was Russian wishing to stay attached to Russia. Then came Mikhail Gorbachev and perestroika and a seriously stagnating Soviet economy, which re-fuelled the separatist independent movements in

Ukraine and many countries in Eastern Europe. On 16 July 1990, the Ukrainians adopted the Declaration of State Sovereignty of Ukraine and following the failure of some communist leaders to be rid of Mikhail Gorbachev, independence was announced on 24 August 1991. In 1994 there was an agreement known as the Budapest memorandum when Ukraine handed over its nuclear weapons to Russia in exchange for security guarantees and territorial integrity from the West and Russia.

There was constant unrest in the Donbas region possibly assisted by Russians in support of the Donetsk People's Republic and the Luhansk People's Republic which led to bitter internal strife. It resulted in a form of frozen stalemate. The problem was the ethnic divisions within this region. The Ukrainian speaking population was about 78 per cent with the Russians having the largest minority of just over 17 per cent, whereas other nationalities, mainly Eastern European have less than one per cent each speaking Russian. There was internal strife over schooling as to where Russian was taught with clashes somewhat escalating in 2009 with the pro-Russian elements. A small Russian organisation known as the Eurasian Youth League organised Russian Marches which soon involved the riot police. There was also a failing demand for unity between the Russian and Ukrainian Orthodox Churches, and the small civil-type conflict was based on an ethnic and political form of nationalism.

For a time, Ukraine stayed loosely connected with the post-communist independent states but in 2013 following mass demonstrations it led to the establishment of a new government and some pro-Russian unrest, and soon unmarked Russian troops occupied and annexed the Crimean Peninsula. In late 2013 the then Ukrainian President dismissed a deal with the European Union to support stronger ties with Russia which led to protests in this divided country.

Meanwhile there was the usual pro-Russian unrest in the predominantly Russian area of Eastern Ukraine, the Donbas area where there was constant conflict between the pro-Russian elements of the population and the government which was seeking closer ties to the West. At the turn of the years 2021-2 it was noted that Russian forces were gathering along the borders, but President Putin asserted it was only a military exercise. In

February 2022 the invasion started which continues as of this moment (December 2022).

This possibility had been simmering since the occupation of Crimea with Russia recognising the separatist regions of Donetsk and Luhansk. The Ukrainian President Volodymyr Zelensky urged his people to resist, and the soggy and difficult Ukrainian terrain made progress at ground level difficult, especially with supply lines. Kiev under missile attack survived and the Ukrainians assisted by modern Western weapons in places drove the Russians back, but the Russian missile system started a process of attacking power plants, and the war in general created a massive refugee problem, with an estimated seven million displaced persons. The facts and the figures, the political machinations, the accusations by both sides, the challenges about war crimes must be left to future historians and not for this chapter.

Many Western governments and publics have offered support to Ukraine, but the potentiality for escalation and the economic decline in the global economy has set in motion in some supporting countries the demand for some form of peace settlement. This chapter was written in December 2022 and only time has the answer. When a missile fell on the Ukrainian-Polish border killing two Poles the Russians denied the attack (mid-November 2022), the Americans and others quickly pointed out that it was probably a defensive Ukrainian missile which went wrong. The Ukrainian government appeared to acknowledge this but blamed the Russians for starting the war in the first place. For many observers this incident seemed to offer some hope that escalation was being avoided, but at the time of writing there remains deep concerns.

The Western powers have imposed massive sanctions on Russia, but Russia was also a major supplier of major resources such as oil and gas to many European nations, and the war has helped provoke a rise in prices at a global level and national economies around the world are suffering with major recessions likely. The third superpower China has proved reticent to criticise Russia by not imposing sanctions and the war has now divided the globe. President Zelensky in many well-tuned meetings and broadcast speeches has managed to gather considerable sympathy in the West while President Vladimir Putin is painted as the villain.

President Zelensky was born in 1978 to a Jewish family and grew up as a native Russian speaker. He obtained an LLB degree but became a comedy actor and in one series played the role of president which was highly popular. He won the real presidential election in 2019 having positioned himself as anti-corruption and anti-establishment. He also wanted unity within the country between Russian and Ukrainian speakers, promising to end the protracted conflict with Russia, but only to see problems escalate. He has proved popular with his broadcasts and interviews across the world with his pleas and demands for help.

President Putin has a very different background, born in 1952 he worked for the KGB foreign intelligence for 16 years rising to the rank of lieutenant-colonel, entered politics in St Petersburg, then Moscow to join President Boris Yeltsin's administration, served briefly as director of the FSB (Federal Security Service) and Secretary of the Security Council. After Yeltsin's resignation he became acting president, then elected and re-elected in 2004, and has been prime minister or president of Russia from 1999 to today. During his first tenure the Russian economy grew as did his popularity. He led the Russian war against the Chechen separatists, oversaw the war against Georgia, authorised intervention in Syria against the Jihadist groups, and having annexed Crimea initiated the war against Ukraine. As the years have passed various observers had noted that the original democratic process established after the collapse of the Soviet Union has appeared to evaporate and replaced by an authoritarian regime, sometimes noted for human rights abuses with the attacks on political opposition.

The Russia-Ukraine war

Since Tsarist days Russia, rightly or wrongly, and as large and powerful as it is, has often deemed itself hemmed in by potential enemies. The two world wars made this feel even more pertinent, and although Russia was a major contributor to victory, after WWII the Soviets faced the Cold War which in many ways was instigated by a clash of ideologies and a fear of Stalin's intentions. The West formed NATO as a protective shield, and for a time the Soviet response was the Warsaw Pact. Following the dissolution of Soviet Russia, the fear of nuclear war felt anachronistic, nevertheless, NATO expanded its membership of many of the once Soviet dominated eastern European countries. This NATO enlargement challenged what might be called the old school in Russia and could be open to some criticism as the Russian fear of being surrounded by enemies was seen by many to be confirmed. The Russians were fully aware that Article Five of NATO stated that an attack against one NATO member would be considered as an attack against all of NATO. Poland, Hungary, Romania, Bulgaria, Slovakia, Latvia, Lithuania, Czech Republic, Estonia, all once part of Russian control since WWII not only had their independence but rapidly joined the Western alliance. For some Russians this was not only an act of betrayal but increased their fear of being surrounded by enemies. Ukraine was regarded as an exceptional country because it had been part of the Russian hegemony long before WWII, notable for its large minority of Russian speakers. Ukraine provided not only agricultural resources but coastline and ports into the Black Sea. President Putin, forever conscious if not nostalgic for the Soviet Union's past felt Russian status was being eroded, pointing out that Ukraine was an inalienable part of Russian history, culture, and even spiritual space.

When the Ukraine made applications to join the EU it was probably irritating to Russia, but joining NATO was deemed a threat to Russian security. Ukraine as noted had given up its nuclear weapons on the understanding of protection from both sides of the great divide, but Putin and others feared that Western nuclear devices would soon find launch places on Russia's boundaries with Ukraine. It was for Putin a serious threat, but in 2022 when the Ukrainian war started NATO had not accepted Ukraine's suggestion of joining NATO, and there were no nuclear weapons in that country. Following the Russian occupation of Crimea had probably increased the sense of Ukrainian nervousness and helped boost the desire for NATO protection. Prior to the war the whole situation was based on pure hypothesis and speculation by politicians on all sides of the divide.

When Russia placed formidable military forces along the borders of Ukraine it was deploying the age-old tactic of preparing an attack under the pretence of military exercise, but this time the crisis was given world-wide attention because of the widespread usage of news coverage. Often countries direct the news coverage and the narrative of war, and the political forces have their own power drive behind the given news, so in

Russia the news is one of self-defence against potential threats, a form of war of prevention argument and therefore based on political speculation. Another theme is supporting Russian residents in Ukraine, what the Russians may see as a war of intervention, while in the West the conflict is regarded as a war of pure aggression.

There are huge question marks about humankind's behaviour which this book has tried to explore, and many aspects come once again into current perspective, not just as history but now on our own doorstep happening as this is being written. The impacts are felt daily with suggestions that Russia has deliberately blown-up undersea supply lines, which some argue could be interpretated as an act of war, though the supplying of weapons by western powers Russia could also be deemed as interference. Sanctions have been started and proved to be a double-edged sword leading to economic crisis points, with this financial impact on everyday life and the fear of escalation leaving the global population as if standing on a cliff's edge.

It has been argued that Ukraine's sovereign rights had been violated under the UN charter, and the invasion is illegal, but as noted this achieves nothing in stopping the war as one country, especially a powerful military one can simply brush international law aside. If the news is to be believed, then the past conventions of the conduct of war are being abused, with claims of torture and massacres and all forms of abuse against human rights. It will take time for many of the facts and accusations to be verified, but the live television broadcasts indicate that missile attacks on non-combatant civilians have happened.

The Russians could claim it is a war of defence, but Ukraine had not attacked them first, nor was it likely they would, and they were not members of NATO. The pre-emptive argument of defence does not hold water as there were no indications that despite the poor relationships between Russia and Ukraine the latter would know it would be virtual suicide to launch an attack on a major nuclear power, which argument also dismisses the idea of a war of prevention. At one stage the arguments for a war of intervention to save the Ukrainian Russian speakers was raised. This perspective could be challenged that either the Ukrainians were doing their best to resolve this problem based on promises once made by the Ukrainian president, or it was the Russian speakers who were creating havoc. Another motive for intervention was provided by Putin based on the claim that Ukraine was led by Nazis. This immediately seemed like a spurious argument because the Ukrainian president was Jewish, and many observers felt Putin made this claim to cause critics to wonder what had been happening in Ukraine by mentioning the unthinkable. It could be argued that Russia attacked out of fear of the encroaching NATO influence, and for some observers this remains credible, and if studied may reduce the sense of bellicosity by both sides.

This war highlights all the problems indicated in Chapter One about the nature of humankind. Despite the horrors of war humanity still creates conflict from motives of territory, wealth, greed, influence, and status; nothing changes under the sun. Nationalism, ethnicity, race, religion, political ideologies, have all added to the witch's poisoned brew as humanity stumbles from one war to the next. Despite all this war has been regarded as immoral or illegal when not in self-defence, and most of humankind would rather not engage in war. Natural law in any of its definitions indicates that killing for any reason other than self-defence is unjustified.

Unfortunately, the critics who state that war is like a recurring natural disaster may well be correct as humankind simply does not or cannot break the habit. Although a nuclear war could mean the total annihilation of life on our shared planet we still engage in international conflict. It is the nuclear threat which has caused the greatest concern as Putin in a forthright statement reminded the world, meaning Ukraine and NATO that nuclear weapons could be deployed. It was issued as an age-old threat of military power to warn Ukraine not to resist, and the West not to interfere. In a NATO Review publication in November 2022, a research analysist (Pierre de Dreuzy) and researcher (Andrea Gill) at the NATO Defence College in Rome wrote an article about the nuclear threats. It was pointed out the threat had a serious destabilising effect on the international security order for the first time since the end of the Cold War. The question was raised whether nuclear weapons have become a weaker form of compellence, which may have ramifications in other conflicts which needs time to test the veracity of this view, as will whether Russia's nuclear coercion will have any effect on NATO's nuclear posture. The article for the reader left no doubt that even a war in one country can suddenly escalate into a global disaster leaving the world standing on a cliff-edge.

Final thoughts

From the famous philosophers to the best religious thinkers, and in nearly every form of morality aggressive war has been condemned. Yet throughout recorded history war has escalated with the horrors of the modern world's advanced technology of weapons of mass destruction, making arguments relating to non-combatants and the innocence totally redundant. Reflecting Georg Hegel, Bernard Shaw once said 'we learn from history that we learn nothing from history'. We have not learnt from the greatest thinkers and intellectuals from the past but confine their offerings to the little-used dusty bookshelves in a few libraries.

However, on these bookshelves maybe the only solution. The world is currently facing its end-days with climate change and concerned that a nuclear war remains a possibility. Seeking a new Utopia where countries can live in harmony and work together to survive, it would make sound common sense if there were to be a gathering of nations, possibly in a setting outside the UN whose main occupation was to discuss the future. Not to sit in ideological or religious groupings, but an open and frank discussion on survival and therefore human conduct. It would demand acknowledging the wrongs and behaviour of the past as that would be the only means of establishing any opportunity for peace and reconstruction for the future. This would oblige agreeing on a universal accepted morality of behaviour (somewhat like natural law) on major issues such as climate change, war, and mutual support. It almost needs a world government, but this is Utopia and all too sadly unlikely.

This book has confirmed the thinking of a fellow priest who told the writer that original sin is here to stay, which is depressing, and for my part, I shall seek my comfort in the Christian faith, and that soon *Jus Post Bellum* may bring peace and harmony which according to St Augustine was the reason for war.

Abbreviations

CN	Campaign for Nuclear Disarmament
CO	Conscientious Objector
EUC	End-user certificate (Arms Sales)
FBI	Federal Bureau of Investigation
FFI	French Forces of the Interior
FSB	Federal Security Service (Russia)
KGB	Committee for State Security (Russia)
ICC	International Criminal Court
IMT	International Military Tribunal
IMTFE	International Military Tribunal for the Far East
IRA	Irish Republican Army
MAD	Mutually Assured Destruction
NASA	National Aeronautics and Space Administration
NATO	North Atlantic Treaty Organisation
OKW	Oberkommando der Wehrmacht (German High Command)
OUN	Organisation of Ukrainian Nationalists
POW	Prisoner of War
ROC	Republic of China
SNP	Subsequent Nuremberg Proceedings

SS Schutzstaffel (Nazi paramilitary organisation)

TRC Truth and Reconciliation Commission

UN United Nations

WMD Weapons of Mass Destruction

Bibliography

Bartov, O., Grossmann, A., and Nolan, A. Crimes of War (New York: The New Press, 2002)

Bellamy, Alex J, Just Wars (Cambridge: Polity Press, 2006)

Burleigh Michael, Moral Combat (London: Harper Press, 2010)

Cameron J, (Ed) The Peleus Trial (London: Hodge, 1948)

Campbell C A. (Professor of Logic and Rhetoric) On Selfhood and Godhood (London: George Allen and Unwin, 1956)

Carey John, Dunlap William, and Pritchard, John R, (Eds) International Humanitarian Law: Origins, (Leiden: Brill, 2003)

Davies, Norman. Europe, A History (London: Pimlico 1977)

Evans, Richard, The Pursuit of Power (London: Penguin Books, 2017)

Fergusson Niall, The War of the World (London: Allen Lane, 2006)

Finney, Patrick, Remembering the Road to World War Two International history, National Identity, Collective Memory (London & New York: Routledge, 2011)

Freeman M.D.A., Lloyd's Introduction to Jurisprudence (Sixth Edition) (London: Sweet & Maxwell, 1994)

Fulbrook Mary (Ed), Twentieth Century Germany (London: Hodder Headline Group, 2001)

Grey of Fakkoden, Twenty-Five Years (London: Hodder and Stoughton) 1926) Vol. 1

Harari, Yuval Noah, Sapiens, A Brief History of Humankind (London: Harvill Secker, 2014)

Hart H. L. A., The Concept of Law (Oxford: Clarendon Press, Second Edition, 1994)

Hastings Max, All Hell Let Loose (London: Harper, 2011)

Hébert, Valerie Geneviève. Hitler's Generals on Trial (Kansas: University Press of Kansas, 2010)

Howard, Michael, The Invention of Peace and the Reinvention of War (London: Profile Books, 2001)

Howard Michael, Clausewitz, A Very Short Introduction (Oxford: OUP, 2002)

Howard, Michael, War and the Liberal Conscience (London: Hurst & Company, 1981)

Kinsella D, Carr C. L. (Eds) The Morality of War, A Reader (London: Lynne Rienner, 2007

Mack Smith, Denis, Benedetto Croce: history and politics, Journal of Contemporary History, vol. 8, no. 1, 1973, pp. 41–61.

Manvell R & Fraenkel H, The Incomparable Crime (London: Heinemann, 1967)

Milgram Stanley, Obedience to Authority (New York: Harper and Row, 1974

Morgenthau Hans, J, Politics Among nations: the Struggle for Power and Peace (New York: Knopf (4th Edition), 1967)

Neave, Airey, Nuremberg, ((London: Biteback Publishing,1978)

Overy, Richard, Interrogations (London: Penguin Books, 2002)

Preston Paul, Franco (London: Fontana Press, 1995)

Reynolds, David, In Command of History: Churchill Fighting and Writing the Second World War (London: Allen Lane, 2004).

Raiber Richard, Anatomy of Perjury (Newark: University of Delaware Press, 2008)

Robertson E. M. (Ed.), The Origins of the Second World War: Historical Interpretations (London: Macmillan, 1971)

Russell, Bertrand Vol 13 (ed. by Rempel, Richard A,) From Prophecy and Dissent, The Collected Papers of Bertrand Russell (London: Unwin House Hyman, 1988)

Sangster, Andrew, Blind Obedience and Denial (Oxford: Casemate, 2022)

Salter, Michael. Nazi War Crimes, US Intelligence and Selective Prosecution at Nuremberg (Abingdon: Routledge Cavendish, 2007)

Salvemini, Gaetano, Prelude to World War II (London: Gollancz, 1953)

Sinclair Andrew, An Anatomy of Terror, A History of Terrorism (London: Pan Books, 2004)

Soviet Information Bureau, Falsifiers of History (Historical Information) (London: Soviet News, 1948)

Taylor, Telford. The Anatomy of the Nuremberg Trials (New York: Alfred Knopf, 1992)

The UN War Crimes Commission, (1949) Law-Reports of Trials of War Criminals, Volume VIII London; Volume VIII. Case No 47, London: Law Reports of Trials of War Criminals

Waltzer, Michael, Just and Unjust Wars, 5th Edition, (New York: Basic Books, 2015)

Index

www.ingramcontent.com/pod-product-compliance
Lightning Source LLC
Chambersburg PA
CBHW021434180326
41458CB00001B/268